THE BIOLOGY OF GRASSES

The Biology of Grasses

G.P. Chapman BSc, PhD, FLS
Department of Biochemistry and Biological Sciences
Wye College
University of London, UK

CAB INTERNATIONAL

CAB INTERNATIONAL   Tel: +44 (0)1491 832111
Wallingford   Fax: +44 (0)1491 833508
Oxon OX10 8DE   E-mail: cabi@cabi.org
UK   Telex: 847964 (COMAGG G)

A catalogue record for this book is available from the British Library.

ISBN 0 85199 111 4

Typeset in Plantin by Techset Composition Ltd, Salisbury
Printed and bound in the UK by Biddles Ltd, Guildford

Contents

v

Preface

Due to unusual weather conditions a male Neolithic human body emerged from a glacier in the Alps in 1991 after having been interred there for about 50 centuries. Among the man's possessions was a cape made from plaited grass stems that could have served too as a portable hide for hunting. His shoe uppers were made from grass and more grass was pushed inside for insulation. His knife was held to his belt by a cord of woven grass. Evidence of his food supply included traces of wheat and, if he were a herdsman, he would have depended on sheep and goats amid these alpine pastures.

Along with animal skin clothing and simple weapons, grasses in one form or another were integral to his way of life.

He is about half as old as agriculture.

Given the multitude of uses to which grasses are now put, their importance is self-evident. Their close association with humankind extends from beyond history and their study is an absorbing challenge.

In three earlier volumes, *Reproductive Versatility in the Grasses*, its companion *Grass Evolution and Domestication*, and another volume *Desertified Grasslands: Their Biology and Management*, I edited the work of some 50 specialists. So as to make some of this material, together with the basic biology of grasses, available to students I wrote with my colleague, Dr W. E. Peat, *An Introduction to the Grasses (Including Bamboos and Cereals)*. Covering broadly the same area, though in a different way, is *Grass ID*, (Dawson *et al.*) a computerised tutorial. For students unfamiliar with grass biology one or other of the introductory volumes should be consulted before working with the present volume.

To edit the work of other specialists is rather like climbing mountain peaks amid a surrounding landscape. Eventually, it provokes questions about that landscape itself and I found myself repeatedly trying to make 'connections'. The present text therefore is an individual perspective and not an all-embracing compendium. What I have written is, in some ways, a reaction to what I have read, as well as an attempt to share the experiences of working with this remarkable group of plants.

Perhaps surprisingly, palaeoagrostology finds a place, as do such strange grasses as *Anomochloa* and *Streptochaeta*. Continental drift is nowadays as much a part of our thinking as evolution has been for rather longer and so I have included a treatment of grasses in the southern hemisphere since the Cretaceous.

If attention were confined to the grasses of Western Europe or even to those of economic importance the result for students would be an over-concentration on relatively similar pooid grasses. My aversion to this in any case was fostered by the arrival, year after year, of overseas students at Wye, combined with the opportunity for considerable foreign travel on my own part. That a world view is so difficult to achieve is not a reason to abandon the attempt and to that end I have taken examples from many parts of the grass family.

I believe it is realistic, with one major exception, to regard the evolution of our principal cereals as essentially resolved. Their treatment here is deliberately brief since I saw no need to repeat what is set out in numerous texts elsewhere. The exception is maize, whose origins remain controversial, thought-provoking and well worth close attention. I make no apology, therefore, though this much of an 'apologia', for the emphasis it receives in the concluding chapter.

Within a generation we have seen the transformation of ecology. The emphasis has changed from one where nature set the pace and students absorbed ideas of orderly succession to one where landscape is torn apart and ecology politicised. There is a sense in which, now, we are all ecologists and this book would be deficient if it ignored such issues – not that it would, since the special properties of grasses relate so directly to environmental disturbance.

Familiarity with grass literature can, curiously, obscure just how strange some of the terms are that we use. It might be simpler if we had terms restricted only to grasses but in practice they overlap, sometimes confusingly, with those for other plant groups. Not just to clarify matters for the student, but because it is worth asking what we really mean, the reader's attention is drawn to the Glossary to which is prefixed 'Critical'. I have even for example, sought to make sense of the term 'raceme' when applied to grasses. This is not to regard definition as a fetish but to use it as one more way of encouraging curiosity about grasses.

One considerable matter remains for comment – namely the bamboos – and throughout I have sought to include them as integral members of the Poaceae which they undoubtedly are. They are both problematic and important. Quite apart from their already major and increasing significance as a renewable (and industrialisable) resource, they have light to shed on the origin and evolution of the family. It is for this reason that there is to be a further volume – *The Bamboos* – edited rather than authored.

Acquaintance with grasses consistently reminds one of their unobtrusive efficiency. When considering their various adaptations everything seems to have been done by fine adjustment. One needs an appreciation of the extent to which sometimes surprising ends are achieved by subtle means, and in any case grasses are an acquired taste. These things being so, I will be pleased if some who have hitherto regarded grasses as baffling, humdrum or merely utilitarian now find them objects of compelling interest and worthy of not a little wonder – for so they are.

G.P. Chapman
Wye College, July 1995

Acknowledgements

I am grateful to many people, and indeed gathering material into a book recalls so much help given so readily. I think particularly of my colleague, Dr W.E. Peat, with whom on an almost daily basis over many years some aspect of grasses has come up for discussion and my too ready assumptions rigorously challenged. At its completion he read the entire book and made various suggestions, and for this I am especially grateful. Additionally, my thanks are due to Professor Dennis Baker and Malcolm Kernick at Wye and to Deryck Clayton, Frances Cook, Tom Cope, Stephen Renvoize, Chris Stapleton and Marylin Ward at Kew, and others I see more rarely, including Professor Ren Jizhou (China), Margaret Friedel, Graham Griffin, Bruce Knox, Peter Latz, Mike Lazarides and Les Watson (Australia). I am indebted to Bernard Peyre de Fabrègues of IEMVT for his help regarding the Sahel.

I thank the librarians at Wye College, Royal Botanic Gardens, Kew, and the Linnean Society of London for their skill and patience in retrieving even the most obscure documents. I am indeed grateful for access to the original drawings in the Kew archive, of which a number are reproduced in these pages. Mr Jeff Brooks at Wye College has, unless otherwise indicated, photographed the grasses presented here and where necessary provided original diagrams.

Shelagh Reardon and Fiona Holt are thanked for the leaf section used in the discussion of photosynthesis, as is my student Nasruddin Aris for the leaf electron photomicrograph of *Dactyloctenium ctenoides* in that section relating to salinisation. I am indebted to the *Journal of Ecology* and Dr M.H. Peart for permission to reproduce his grass diaspore drawings.

My thanks are due to the Institute of Archaeology, London, for access to the Jericho archive and to the Linnean Society of London for permission to photograph Linnaeus' original maize specimen.

Alan Woods, a former student at Wye, brought to my attention the work of his father, Robin Woods, regarding *Poa flabellata* in the Falkland Islands and I thank them respectively for advice and permission to reproduce items. To Wang Kanglin I am indebted for details of bamboo taxonomy in the forests of south-west Yunnan and to Dr R.S. Nadgauda for her detailed observations on bamboo stigmatic secretions. I thank Professor Elizabeth Trusswell for advice on Australian geology, Professor David White for his observations on *Phragmites* in the Mississippi delta, Dr Surrey Jacobs for guidance on Australian spinifex and Dr Gerry Hooker for help with the horse fossil lineage. Even so, I accept that the shortcomings of this book are my own.

It is a pleasure to thank Miss Pam Kite who, over many years, has grown a long list of grass species for me.

There are two groups of people almost invariably overlooked among acknowledgements and yet each is essential to the kind of work detailed here. These are the drivers whose skill and cheerfulness have sustained me across the most formidable terrain and the translators who made conversation possible. To each I express my gratitude.

To the Royal Society and the British Council officers I, like many others, owe much to their skill in both making international contacts and financing them.

Margaret Critchley and Jocelyn Hart have, with patience and good humour, word-processed the entire text. My thanks are due to Sue Briant, Brenda Ladley and Laura Sessions for much valued additional help typing indexes. I thank the staff of CABI for help throughout.

Finally I thank my wife, Sheila, who remains my patient, witty and understanding companion.

A Note on Illustration

As commonly regarded, grasses are thought of as uniformly green. Closer acquaintance shows them, as a family, to have the entire spectrum of colours and presented in such a way as to test the skill of the most expert watercolourist. This is especially so if botanical detail is to be preserved rather than creating something merely impressionistic.

Grass photography, too, is demanding. What to the ecologist or taxonomist, in the field, is so readily obvious can prove next to impossible to capture in black and white. Even when attention is confined to individual plants or inflorescences subtle variations in shade make considerable demands on the photographer. Matters have certainly improved with the advent of colour photography, though not perhaps as much as might be expected for grassy landscapes. The human eye can take in movement and differential focus in a way that often shows up the inadequacies of colour photography. And, are grass experts necessarily photographers above the ordinary? It is at the level of spikelet and floral close-up that colour pictures are more often genuinely satisfactory. The combination of close focus and variation in colour and texture, together of course with precisely controlled lighting, lie behind what comprises a successful grass photograph. None the less, reproducing colour photography in print is expensive and, without extreme care, the contrast between the original colour slide and the result on the page can be all too evident to the originator.

Agrostologists along with other botanists, for whom fine detail is crucial, have followed a solution to these problems which might seem surprising. This is the pen and ink drawing. Its advantages are that it can show what the author wishes in terms of proportion and fine detail, leaving out what is extraneous. When printed, such accuracy can be conveyed far more cheaply than by colour photography and in many cases more effectively. A more or less standard format is to allow one page to indicate both the whole plant aspect and its component parts while concentrating on features that matter to the taxonomist.

There seems little likelihood in the near future that some kind of computerised system will substitute convincingly for the work of the traditional botanical illustrator. At the same time it does require practice to 'read' such drawings. For these reasons substantial use has been made of such illustrations in the present book. I have, for example, sometimes included a botanical drawing of a particular species when the matter under discussion is not necessarily taxonomic.

In this part of the work I have been greatly helped by the staff at the Royal Botanic Gardens, Kew, mentioned earlier. The drawings of particular species here represent a selection of those with which botanists work and, mostly, were published first in either the *Kew Bulletin* or in various regional Floras. They are as follows: D. Erasmus, *Arundo donax*, *Chloris gayana*, *Leptochloa fusca*, *Setaria glauca*, *Stipa capensis*, *Stipagrostis ciliata* and *Zea mays*; J.C. Erasmus (formerly Webb), *Alopecurus myosuroides*, *Cynodon dactylon*, *Hyparrhenia hirta*, *Panicum turgidum* and *Tripogon minimus*; M. Grierson, *Festuca killickii* and *Nastus hooglandii*; L. Ripley, *Hygrorhiza aristata*; S. Ross-Craig, *Centotheca lappacea*; and A. Webster, *Sinarundinaria alpina*.

Grasses in a Changing World

1

The diversity of the world's living matter is now reduced to order. Linnaeus and his successors have provided a serviceable framework that has both survived Darwin and helped sharpen the evolutionary questions we need to ask. Similarity we now interpret, where we can, as relationship. Contemporary microspeciation we handle experimentally, assuming it to be a continuation of the processes which, earlier, delineated what became families and genera.

What we find, we have come to realise, does not quite fit our requirements and using nature's means, genetics applied to plant breeding, we have set about making improvements. Latterly, we have gone a stage further, almost it seems disparaging nature's means, setting out to fashion new ones. We have hybridised biology with ballistics to make 'biolistics'. No matter that one of our cereals is deficient in this or that gene, we propose to add it anyway. If we are not yet masters of the situation it appears that we intend to be and our crop plants, grasses among them, are as clay to be modelled. But, how like one grass is another? What diversity of form and function already exists? Could a wider knowledge of grasses open up new possibilities? If this family, already global in its distribution, is undeniably an ecological success, what is the basis of that success? And, if some parts of the grass family seem more suitable to us, is it because there is there some inherent malleability or responsiveness with which we can work more easily? To what does a careful assessment of the grass family amount?

Interdependence

Two families, one animal, one plant, which are to a considerable degree interdependent, are, also, especially conspicuous. One is the Hominidae with ourselves as its only surviving species. The other is the Poaceae with about 10,000 species and which provides cereals, forage and cane sugar. It includes, too, the bamboos, which have an immense variety of uses.

To understand the interdependence requires, on the one side, the study in its broadest sense of anthropology and on the other an understanding of the special features that predispose grasses to domestication. Grasses present the biologist with a plant body and a reproductive mechanism each deceptively simple but remarkably adapted to survival amid the colossal disturbance wrought by our own species across the earth.

Long before *Homo sapiens* became conspicuous, grasses had and were evolving their distinctive features. With the advent of agriculture late in human evolution and very late for grasses, domestication marked a new direction. The story of grasses has therefore recognisably pre- and post-agricultural phases, each of which needs to be understood.

The 'Distinctiveness' of Grasses

With their linear leaves and small green flowers, grasses are readily recognised and confusion with other families is unlikely except for rushes (Juncaceae), sedges (Cyperaceae) or more rarely with an obscure group of southern hemisphere families that includes Anarthriaceae, Centrolepidaceae, Ecdeiocoleaceae, Flagellariaceae, Joinvilleaceae and Restionaceae (Dahlgren *et al.*, 1985; Watson, 1990). Even woody bamboos many metres tall appear 'grassy' while, at the same time, extending our notion of grasses.

Characteristic features of grasses include a fruit containing one seed that has a generous food reserve and a well-differentiated embryo at maturity. At germination, their combined effect is to give the plant a 'running start'. The organism that develops is seldom woody, nor, it would seem, highly elaborate, and onset of the next reproductive phase can be relatively rapid. Again, annual and perennial grasses tend to have numerous growth points, which, whether they develop into branches, rhizomes, stolons or tillers, contribute to a highly competitive growth habit. Grasses are generally drought tolerant, and if they brown in dry weather rapidly regreen and resume growth with the return of moist conditions.

Grass Habitats

The features just outlined combine to fit grasses for open habitats so that they tend to occur as pioneer or early colonists. Where neither temperature nor moisture are limiting, grasses are eventually overgrown by shrubs and trees and then comprise a less significant part of the vegetation. Here, as in other respects, bamboos are exceptional and may develop amid the forest canopy as a scrambling growth form like *Chusquea* or even produce a dominant tree-like climax.

Where moisture is limiting for tree growth grasses can provide the dominant vegetation as in tropical savanna or even some of the isolated clumps of plant life found in deserts.

Bamboos

A recurrent theme in grass biology is an attempt to generalise while pointing to the bamboos as exceptional. A few examples will suffice. In addition to woodiness, bamboos differ from most grasses in their 'trimerous' flowers, a subject explored in detail later. Again, bamboos in some cases have long time intervals, perhaps as much as a century, between germination and flowering. Indeed, they manifest 'mast' flowering where similar bamboos under different circumstances flower at nearly the same time as if prompted by some almost imperturbable inner clock.

The perception that bamboos are somehow odd is reinforced by the fact that most agrostologists work on what they tend to call 'true grasses', by which they include cereals, forage and lawn grasses. Had they considered the matter they might, about as usefully, have referred to 'non-bamboos'. A consequence is that bamboos have come to seem 'oriental' and 'exotic', both botanically and in the more lax general uses of these words. Yet grasses they unmistakably are, and somehow we have to accommodate them or else distort our science. To omit them is to have only a partial view of the family and to include them is to load ourselves with questions that might seem awkward and inconvenient at first but which properly extend our understanding.

Are bamboos primitive because of their more lily-like flowers or more advanced through their massive woody culms? Could they be both at the same time? Are they quaint archaic survivors or are they at the base of 'mainstream' grass evolution? One difficulty for many botanists, both teachers and taught, has been that bamboos were inaccessible and shy flowering. Given modern travel and transport and imaginative planting it is indefensible now to exclude them from a well stocked collection of important and instructive plants. Watson (1990a) remarked:

> Clearly people trying to discover in grasses, or wanting to generalise about the family from minimal observations, should deliberately seek to maximise the diversity of their samples by arranging to include (say) a bamboo (or rice), wheat and maize.

There is a widely perceived association of China with bamboos and certainly the southern half of China is particularly rich in species, quite apart from the varied uses to which they are put, but this is too narrow a view.

Fig. 1.1. *Gigantochloa laevis* used as scaffolding for a multistorey building in Guanzhou (Canton) south China.

Even to extend the association to India and south-east Asia is inadequate. There exist in the New World tropics a group of bamboos, interesting in their own right, in some ways different from those in the Far East and currently prompting new questions (Clark, 1990). The matter does not stop even here, since, given continental drift, how is there a relative dearth of bamboos in Africa, a continent once contiguous with both occidental and oriental land masses?

Bamboos belong naturally in an inclusive study of grasses and are, therefore, integral to the intentions of this book.

Reproduction

Given its numerous growth points, a perennial grass is well adapted to vegetative reproduction. As with many other angiosperms, grasses demonstrate the evolution of annual or ephemeral forms from perennial forebears, and here the changeover is an interesting one. Many annuals, wheat and barley among them, retain 'tillering' ability and a tiller with its adventitious roots can be separated from the mother plant and propagated independently. What conditions annualism is a massive commitment to seed production, sufficient to senesce the plant. The situation is not irreversible and

Fig. 1.2. *Phleum arenarium*, a diminutive grass growing on sand dunes in south-west England.

genes from *Agropyron* introduced into wheat can create a perennial habit – a situation in which commercial suppliers of wheat seed, of course, see no interest.

Grasses reproduce sexually with a Polygonum-type embryo sac, described later, a feature they share with the majority of flowering plants. Three reproductive features, however, tend to set the grasses apart. These are a two-gene (S–Z) self-incompatibility system apparently unique to the Poaceae, the widespread occurrence of apomixis and the existence of that most remarkable structure the 'spikelet'.

Photosynthesis

At middle latitudes, in Italy for example, it is commonplace to see two highly successful grasses, perennial ryegrass (*Lolium perenne*) and Bermuda grass (*Cynodon dactylon*), co-exist. In winter the balance of advantage lies with the former but shifts to the latter in high summer – a fact readily apparent since the *Lolium* goes brown in summer amid blue–green steadily growing *Cynodon*. Perhaps one's first reaction might be that given global warming the advantage to *Cynodon* would further increase, but the matter deserves much closer scrutiny. *Lolium* has a C_3 and *Cynodon* a C_4 photosynthesis, of which several variants exist. The emergence of C_4

photosynthesis seems likely to be due to a dearth of CO_2 in Tertiary atmospheres (Cerling *et al.*, 1993); for this and other reasons C_3 and C_4 systems are of considerable significance and will be discussed in detail.

Herbivory

Among those animal lineages for which we have a fossil record, that of the horse is remarkably comprehensive. The expansion of an environment favourable to grasses during the Tertiary period evidently provided an opportunity the ancestors of the modern horse could exploit. Horse adaptation to open country by dentition, gait, visual capacity and digestion make it, *par excellence*, a 'grass-eating animal'. The specialised digestive systems of cows, goats, rabbits and sheep similarly indicate a kind of natural harmony between diet and the availability of a convenient food – grass in abundance.

Is the harmony always so obvious? How is it that a stocky bear-like creature so obviously a carnivore by teeth, temperament and the shortness of its digestive tract should slowly or suddenly opt for herbivory? While, figuratively, in another context, we might accept that 'the lion shall eat straw like the ox', it is a quite different matter for a biologist to explain how a panda can stomach bamboo. We are used, of course, to illustrations of pandas munching bamboo but it is none the less a paradox. Not only is its preferred grass high in silica and low in digestibility, but also it persists in neglecting far more nutritious plants that grow in bamboo forest. Is the conservationist's main problem with the panda its perverse commitment to vegetarianism that jeopardises its fertility, security and vigour all at the same time?

What of the grasses themselves? Does herbivory change grasses or is accidental pre-adaptation to herbivory one more element of their success? More realistically, how is it that some grasses are more adapted to herbivory than others?

Secondary Metabolites

Among plant families, the Poaceae, as providers of edible material for a wide variety of herbivores, are not laden with conspicuous amounts of pharmaceutic substances. This is not to say that grasses do not figure in folk medicine because they clearly do and all kinds of remedial claims are advanced on their behalf from snake-bite treatments to anti-microbial activity. A degree of scepticism is appropriate.

None the less, a range of interesting chemicals is known from among grasses of which the most familiar are aromatic oils from the genera

Cymbopogon and *Vetiveria*. A separate group of compounds are those deriving from fungal infections and the deliberate culture of wheat infected with ergot provides a source of medically active compounds.

Perhaps the most interesting chemical found recently to occur in grasses is 6-methoxy benzoxazolanone. This occurs in bamboo shoots, is known to trigger reproduction in rodents and raises the question as to whether it could be similarly important for pandas (Schaller, 1993).

Domestication

We are latecomers. Humankind is a parvenu only about one million years old. Grasses have probably existed for about 70 times as long. Endlessly inventive, even at so great an age, they had the means to respond to us. And so cereal agriculture arose. The statement 'Man domesticated grasses and thus evolved cereals' can be challenged on two fronts. Firstly, in some traditional societies it is evident that women care for the crops, not merely their cultivation but also the disposal of them through local markets. In a society when men hunted, later perhaps modified to pastoralism, women in a more sedentary position looking after children might well have attended to the care of seedlings and other stages in the life cycle of a newly culti-vated plant. A second, altogether more philosophical approach, is that of Chaloner (1984) who examined the idea that grasses 'took the initiative' in domesticating us.

Traditionally, the origins of agriculture were clothed in myth and folklore – a veil that has now been penetrated by a combination of archaeology, genetics and taxonomy. From beginnings in the Near East, Central America and China agriculture spread out and became a way of life for much of the human race – a process confined to about the last 10,000 years.

Weediness

The dawn of agriculture meant not only that preferred plants were recog-nised but, additionally, others were brought into cultivation. With them came unbidden and unwelcome the 'camp followers', the weeds that flourished under cultivation and diminish crop productivity. Since weeds survive independently of having to be planted and only fed and watered incidentally, they create an impediment to agriculture. So high a proportion of weeds are grasses that the adaptation of the Poaceae to disturbed ground is there for all to see. The coming of agriculture in this particular way did not create a new situation but vastly extended one which grasses had already long exploited.

Even so, not all grasses are weeds and among those that are the basis is not readily apparent. Weediness is, seemingly, likely to be found almost anywhere in the grass family and it does not follow that because one species is weedy, so will be its close relatives. One exception to this perhaps is the subtribe Setariinae where seven genera have weedy species of which three are of major importance, but even here the remaining 60 or so genera are not significantly weedy. Among the ten grass weeds considered to be of greatest importance only one genus (*Echinochloa*) has contributed more than one species, namely *E. colona* and *E. crus-galli* (Chapman and Peat, 1992). The implication is not to generalise although in a slightly different context there is an interesting exception.

Parsons (1970) examined the removal of forest cover in the lowlands of tropical America through agriculture – at first arable and then replaced with a grass cover. Following either deliberate or accidental introduction from Africa into the New World it is evident that much of the grass cover is of African rather than American origin – the so-called 'Africanisation' of the New World tropical grasslands. Conspicuous examples are *Brachiaria mutica* (paragrass), *Cynodon dactylon* (Bermuda grass), *Digitaria decumbens* (pangola grass), *Hyparrhenia rufa* (jaragua), *Melinis minutiflora* (melado), *Pennisetum clandestinum* (kikuyu grass) and *Panicum maximum* (Guinea grass). Parsons (*ibid.*) writes:

> Although originally they appear to have had quite restricted distributions in Africa, they are today found over wide areas of 'derived' savanna surface that once supported trees. Introduced into America, these grasses have proven to be explosively aggressive, colonising and holding vast areas wherever they have received even minimal support by man.

There are few comparable examples of migration in the opposite direction followed by such emphatic occupancy. In no sense weedy, because of its total dependence on humankind, *Zea mays* (Indian corn, maize) is a conspicuous instance of a New World grass becoming widely adopted beyond the Americas.

The grasses of Australia are, even more than those of the New World, 'inwardly contained'. None appear to have achieved weedy status elsewhere and the indigenous grass flora appears highly specialised. Later, we examine how far these seemingly regional or continental features might be explained.

Urbanisation

Settled agriculture and urbanisation are closely linked. An assured food supply allows those who would rather not farm develop other skills and the rudiments of village and town life, artisans, administrators and some kind of leadership form like crystals and almost as inevitably. Initially, since primitive technology could not invariably guarantee food supplies, both the need to store and periodic shortage were recurrent features of agriculture and it is only within the last 40 years, in the favoured parts of Western civilisation, that the folk memory of food shortage has finally begun to depart. Elsewhere in the world, it remains a perennial problem.

Urbanisation on the scale we now experience, where most of the world's population lives in cities, is a situation largely brought about by the Industrial Revolution since factories with access to power supplies and markets initially required a concentration of cheap labour. Accumulating wealth was diverted partly toward science since it was perceived that new ideas could be directed toward the creation of further wealth. Some such ideas began to involve agriculture and the founding of Rothamsted, the world's first agricultural research station, in 1843 marks the beginning of a commitment to a scientific approach. The application of science in this way, replicated around the world, has brought gains in productivity through control of disease, a better understanding of plant physiology, and genetic improvement through plant breeding. These things in turn have both helped raise expectations and underpin gains in modern medicine. We therefore confront a situation where more people survive to maturity and in better health and population has risen with increasing speed. One response, born out of Western technology has been the 'Green Revolution', undeniably a scientific accomplishment, but occurring often amid political and social structures ill-placed to accommodate it. The further scientific promise of an extended Green Revolution remains and the administrative task is two-fold – firstly to provide the resources to sustain the scientific thrust and secondly to ensure the equitable dispersal of the ensuing benefits.

The Ecological Imperative

The cereal scientist shares with others in agriculture the commitment to raise productivity. At the same time it is recognised, perhaps reluctantly, in the Western world that this is only feasible in the long term within an ecologically sound framework. There is an identifiable common cause between what, to a biologist, is common sense and to a politician 'green' issues. The Westerner expects to have adequate food and a range of choice. Since he or she has money to spend, the energy cost of bringing foreign or out of season foods from far away places is hardly a serious worry for most people and, at the same time, access to a traditional countryside is taken for granted. An underlying assumption is that abundance and amenity can, and should, co-exist sometimes with an unthinking disregard for consequences elsewhere.

For the rest of the world a different framework is in place. Often, at least seasonally, food is in short supply or, given rising population, is likely to be so in the future. The primary drive is for increased or sustained production, poverty is commonplace and shorter-term decisions prevail over longer-term ones. An international bauxite company, for example, offered help to its tenant farmers in the form of free fertiliser to raise their banana yields and so increase their income. The response, in some cases, was simply to sell the fertiliser on the basis that 'money now' was the priority. Under such circumstances amenity and conservation can seem completely irrelevant, if indeed, the farmer and his family have even heard of them. One influence that can be brought to bear is tourism and in many countries things considered of only incidental or trivial consequence begin to acquire quite different status. The African elephant or the Indian tiger hitherto seen as a nuisance or a danger becomes recognised as a national asset since it draws tourist income.

The biologist, quite properly, in this situation is pulled in two directions. The need for increased food is obvious while the ecological damage that can and often does result is readily apparent. That tension so evident to the biologist is one that, professionally, he or she requires to resolve. Higher yields obtained, where possible genetically, rather than by chemical means are preferred and where chemicals must be used they should be the most benign available and applied at the lowest feasible level. The wilderness matters crucially. Not everything is for exploitation and part of the biologist's role is both to be aware of the aesthetic dimension in nature and to be able to communicate it to others.

It will be alleged that the wilderness matters because the biosphere must be seen as a whole. We inherit a functioning interrelated system whose dynamics we should hesitate to redirect and I agree. Again, and with this too, I agree, the wilderness could yield for us new drugs, new and more selective pesticides. Each is in its way an appeal to self-interest neither of which I would dismiss. I believe though, additionally, that the biosphere and the planet it envelops have, quite simply, to be valued for themselves.

Plant Breeding and Molecular Biology

Plant breeding has made and will continue to make a steady contribution to increased yields. It is an art as well as a science and not all its successes can be reduced to rational explanation. Much of it, though not all, requires the steady application of formal genetics and latterly some input from tissue culture.

Genetic recombination mediated through meiosis is only approximately predictable, the generation interval, even in a cereal, cannot be reduced below about four months and in any case even the best equipped controlled environment rooms provide a space limit on the recovery of possibly useful recombinants. It is therefore an attractive prospect if genetic change could be precisely *directed* rather than having to wait upon the uncertain outcome of recombination. The most efficient plant transformation is that involving 'crown gall disease' caused by *Agrobacterium tumefaciens*. Although the system can be made to work with dicotyledons, its application to monocotyledons has been disappointing. To what extent are alternative systems available with stable transformation leading to generation to generation transfer? Subsequently, we examine the accomplishments of classical plant breeding and the prospects for alternative or complementary systems.

Amenity

At lower latitudes the grasses used for open spaces such as lawns and public parks include *Axonopus compressus*, *Cynodon dactylon*, *Stenotaphrum secundatum* and *Zoysia japonica*, *matrella* and *tenuifolia*. Only minor changes might be expected in the foreseeable future to this general pattern.

Further from the equator the common lawn grasses include *Agrostis canina*, *stolonifera* and *tenuis*, *Festuca rubra* and *Poa pratensis*. Traditionally *Lolium perenne* added to a lawn grass-seed mixture allows for harder wear. Since the English climate provides sufficient moisture for the growth of woodland, removal of the tree cover creates an environment sufficiently wet to keep grass green most of the year. If, to this, one adds the national enthusiasm for gardening, and a certain characteristic instinct for convention the preoccupation with 'smart' lawns is to some extent explained. Given a fine leaf grass mixture, a level area and a lawn mower in good condition, mowing in first one and then the other direction creates the quintessential 'pyjama'-striped effect seen on television at Wembley and Wimbledon. This too is unlikely to change but the newer fashion elsewhere for turning lawns into 'meadows' in the interests of wild flower conservation is a welcome development. There is room there not only for a range of colourful dicotyledons but a wider range of grasses among them *Briza media* (quaking grass) and *Lagurus ovatus* (hare's tail).

A further development has been the recognition by nurserymen that specimen single plants of grasses add interest to a garden. *Deschampsia caerulea*, *Festuca ovina* in its grey-leafed form, *Molinia caerulea* and various species of *Stipa* are among those used in this way. Yet more impressively there has been an awakening interest in ornamental bamboo and to the list published by Chao (1989) the activities of enthusiasts have added many more. This trend now well established in the United Kingdom parallels those in North America and Western Europe.

For a variety of reasons, therefore, the range of grasses, normally wide in tropical countries, is accompanied by increasing diversity in temperate regions with implications for research and teaching.

Global Damage

Rising population, of course, increases demand upon resources, prominent among them water and cultivable land. Since the situation is an urgent one it has seldom been well planned and the traveller is insistently confronted with the scars of poor stewardship. These include deforested hillsides, eroded landscapes, salt-poisoned fields and fragile ecologies desertified more by ourselves than by climatic change. A recent estimate (Brown and Flavin, 1988) suggests that 24 billion tons of topsoil are lost each year, coinciding with the addition of a further 88 million people. The deprivation resulting is a breeding ground for unrest, agitation and war and the consequent upheaval further compounds the problem. Against this sombre background

Fig. 1.3. An example of erosion in Mexico. The background shows the earlier surface covered with trees. Clearance and cultivation followed by torrential rain removed soil to a depth of 2–3 m except for columns of more resistant material which were then further eroded by wind and rain. The flat lower surface shows recolonisation by pioneer grasses.

the plant scientist cannot do everything but he or she does have grasses – plants that thrive conspicuously upon ecological disturbance – and it is worth noting that the world's grain harvest has been more than doubled in the last 40 years. And, who knows, perhaps humankind will come to its senses. Given a stabilised population and our best science, a contented people could enjoy the earth's bounty. Whatever we might hope for the future though, the present reality is a degraded environment and a mostly uncaring people. This is the situation we have to address. It is in its way as daunting as putting a man on the moon and a great deal more important.

Since this is the situation in which plant science finds itself it is essential to recognise the fact. Not only does it underline the relevance of our work, it sustains in us a proper sense of urgency. And, too, we are fortunate because in doing what is useful and necessary we can work with plants that have evolved so remarkably.

Grass Diversity 2

The Poaceae is a worldwide family that has entered almost every land habitat. In such circumstances great morphological diversity might be expected but the reality is rather different. What one observes is a surprising degree of seeming uniformity which none the less embodies many subtle adaptations. Some diversity is morphological but the impression of an 'all-purpose plant body' is an insistent one. Closer enquiry shows many physiological differences, of which those for photosynthesis are important both immediately to the plant, but also to our understanding of grass evolution. Among the grasses everything seems to be done by fine adjustment and, often, with surprising results. It is this awareness that helps us understand the significance of grass diversity.

Subfamilies of the Poaceae

The grass family is divided into five subfamilies (Bambusoideae, Arundinoideae, Pooideae, Chloridoideae and Panicoideae) although, additionally, some authorities recognise a small sixth one, the Centothecoideae. The five principal subfamilies have a status widely accepted by taxonomists although, of course, there remain disagreements and uncertainties at the tribe, genus and species levels. Two systems of grass classification are widely current. These are Clayton and Renvoize (1986) *Genera Graminum* of which a summary can be found, with a small degree of updating, in Clayton and Renvoize (1992), and the other system is that due to Watson and Dallwitz (1988). A summary can be found in Watson (1990b) and the complete version in *The Grass Genera of the World* (Watson and Dallwitz, 1992). The two summaries mentioned are set out in similar format and permit convenient comparison. Later, these two alternative taxonomies will be discussed in some detail. For the moment, the need is to introduce the subfamilies so as to provide a systematic overview of grass diversity. (CR and WD signify the alternative taxonomic viewpoints.)

Although a fuller treatment of grass taxonomy is reserved for Chapter 6, the material presented here is by subfamilies because they provide a convenient framework with which the reader will find it useful to become familiar.

1. Bambusoideae (CR 13 tribes, WD 2 supertribes, 15 tribes)

Mostly perennial, culms woody or herbaceous, these include the most visually impressive grasses. Some, like *Dendrocalamus*, are immense (rising to 40 m) but many are relatively diminutive and *Arundinaria disticha*, although woody, forms almost a lawn. Among herbaceous representatives are the rice group that includes *Oryza sativa* and American wild rice, *Zizania aquatica*.

Technically important features of this subfamily include the following – leaf blades often pseudopetiolate and disarticulating, leaf in transverse section showing arm and/or fusoid cells and with uniformly C_3 type anatomy. Flowers such as occur in the woody bamboos are often trimerous. Rice is arguably more modern with two lodicules and stigmas but 'retaining' six anthers. The basic chromosome number is $x = 10$, 11 or 12.

Culms of *Gigantochloa laevis* are used to form scaffolding for high-rise buildings and are said to be more reliable in a typhoon than using steel (see Fig. 1.1). Clumps of *Bambusa vulgaris* are found throughout the moist tropics and provide both some defence against soil erosion and a commonly used supply of support canes for gardening there. Such plants conform to our most common perception of what bamboos are like. It is however only a partial view and among the Bambusoideae there is a far greater diversity to be found. *Chusquea* is hardly 'stately' but has instead a scrambling growth among tropical forest, as does *Nastus*. And if, almost automatically, one associates bamboos with hot countries then it is salutary to recognise that some *Thamnocalamus* species survive in the understorey of pine forest in Tibet at about 3000 m above sea level with temperatures of −20 or −30°C for five to six months of the year. By contrast, the genus *Pariana* inhabits the understorey of tropical forest. This last has 6–30 conspicuous yellow anthers per floret and is insect pollinated, a most unusual departure for bamboos or indeed the grass family and examined further in Chapter 4. Three representatives are illustrated here: *Nastus hooglandii* (Fig. 2.1), *Sinarundinaria alpina* (Fig. 2.2) and a curious member of the Bambusoideae, the free-floating genus *Hygrorhiza* (Fig. 2.3).

Because bamboos are, normally, vegetatively propagated, all manner of chimaera will survive and over many years have provided an array of horticulturally useful plants. These include *Arundinaria viridistriata*, with yellow and green striped leaves, a condition that occurs in many other

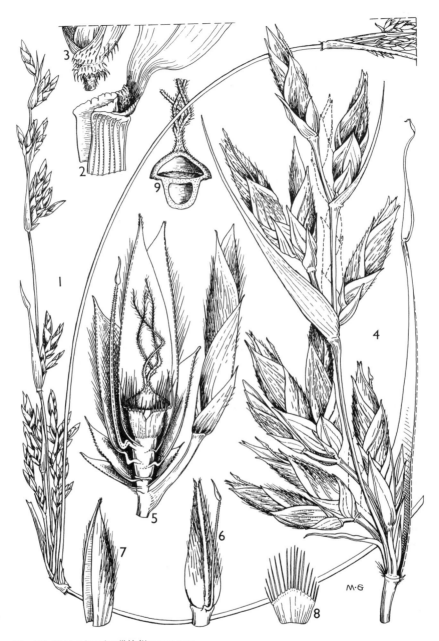

Fig. 2.1. *Nastus hooglandii* Holttum *sp. nov.*
1. habit × $\frac{2}{3}$; **2.** top of leaf sheath with ligule and base of blade × 5.7; **3.** pulvinus and bases of sheaths showing stiff reflexed hairs × 2.8; **4.** part of panicle with half sheath removed to show base × 2.8; **5.** spikelet in sectional view with no stamens present × 5.7; **6.** palea with extension of rachilla and rudimentary floret × 3.8; **7.** lemma × 3.8; **8.** lodicule × 5.7; **9.** ovary, longitudinal section × 7.6.
A climbing or trailing bamboo with spikelets having about five glumes. Occurs in New Guinea and considered by Holttum (1967) to be similar to *N. borbonicus* of Madagascar.

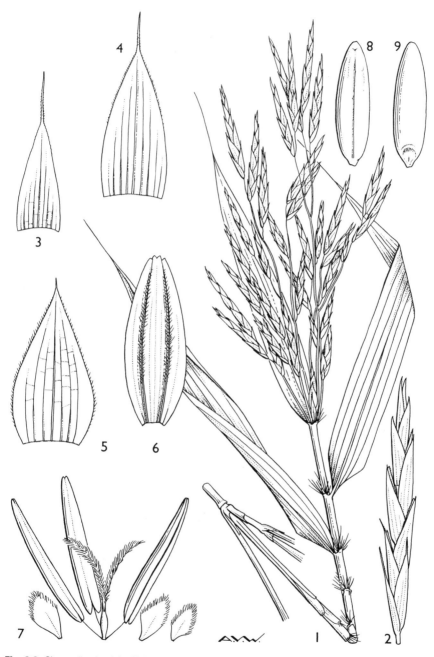

Fig. 2.2. *Sinarundinaria alpina* Nakai.
1. panicle × 1; **2.** spikelet × 4.7; **3.** lower glume × 6.6; **4.** upper glume × 6.6; **5.** lemma × 5.6; **6.** palea × 5.6; **7.** floret × 9.5; **8/9.** caryopsis × 5.6

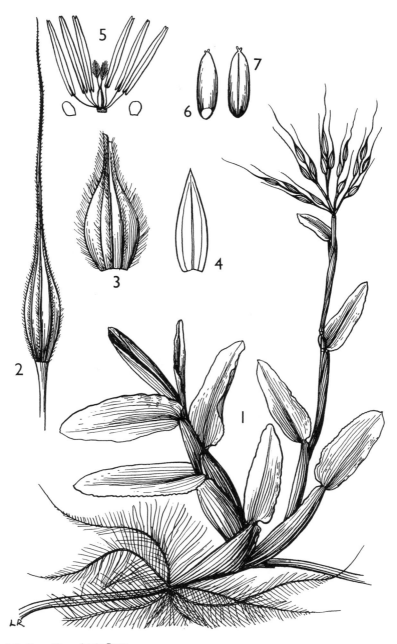

Fig. 2.3. *Hygrorhiza aristata* Desv.
1. Plant × 1; **2.** spikelet having no glumes; **3.** lemma (awn not shown); **4.** palea; **5.** flower including stamens, ovary and lodicules; **6.** grain showing embryo; **7.** grain showing linear hilum. All × 8.
H. aristata occurs in free-floating masses in lakes and slow-moving streams and is eaten by cattle. Poorer people harvest the grains. Cultivated in Assam (Bor, 1960).

bamboos. *Bambusa ventricosa*, with bulbous internodes, is the 'Buddha's belly' bamboo. (By contrast, a variant of *Phyllostachys aurea* bulges at the top of each internode, a condition recalling the human skull.) *Phyllostachys bambusoides* has among its numerous variants 'Castellonii'. Here a given node is mostly yellow but with a green vertical stripe. The nodes above and below are coloured similarly but the green stripe is on the opposite side of the culm.

In China about half the bamboo forest is *Phyllostachys edulis*, the principal source of edible bamboo shoots. It has produced several variants including 'tortoiseshell bamboo' where alternate nodes bulge in opposite directions. The curious plant *Chimonobambusa quadrangularis* produces stems square in cross-section.

Among the strangest, not only of the bamboos but of the entire grass family, is *Anomochloa*, a genus discussed later in some detail.

2. Arundinoideae (CR 3 tribes, WD 11 tribes)

Key features include – ligule a fringe of hairs, inflorescence a panicle, spikelets one to several flowered, laterally compressed, disarticulating above the glumes, florets nearest the glumes sometimes incomplete. The leaf in transverse section shows C_3-type anatomy or in the case of *Aristida* C_4 (NADP-ME). These and other C_4 variants are discussed in Chapter 8. Basic chromosome number is typically $x = 12$ although instances of $x = 9$ and 7 also occur.

Plants in this subfamily vary in size from diminutive herbs to such genera as *Arundo* sometimes 3 or 4 m tall though with a less mechanically robust cane than is found among bamboos (Fig. 2.4).

Arundinoideae are commoner in the southern hemisphere where, it is alleged, they show 'relict' status. Renvoize and Clayton (1992) regard them as the 'descendant of an ancestral line closest to the earliest true grasses'. *Aristida* on this reckoning with its 300 or so species seems like an offshoot with a new lease of life and, by means of its C_4 photosynthesis, very successfully exploiting arid and semi-arid habitats. No cereals occur in the Arundinoideae but several well-known or useful grasses belong here. These include *Cortaderia seloana*, the familiar ornamental 'pampas grass' and *Phragmites australis*, the common 'reed' used for thatching and screens. Less familiar are *Stipa barbata*, an ornamental with spectacular awns, and *Thysanolaena maxima*, an impressive grass found for example in moist tropical clearings. It is worth noting that the tribe Stipeae, which includes *Stipa* although placed in the subfamily Arundinoideae by Watson and Dallwitz, is considered by Clayton and Renvoize to belong to the Pooideae. A remarkable member of this subfamily is the genus *Micraira*, having, in Australia, a group of 'resurrection' species that rapidly regreen after rainfall. *Micraira* is almost unique among grasses having a spiral phyllotaxy, the only other known instance being *Arundoclaytonia*, an unrelated genus in the subfamily Panicoideae.

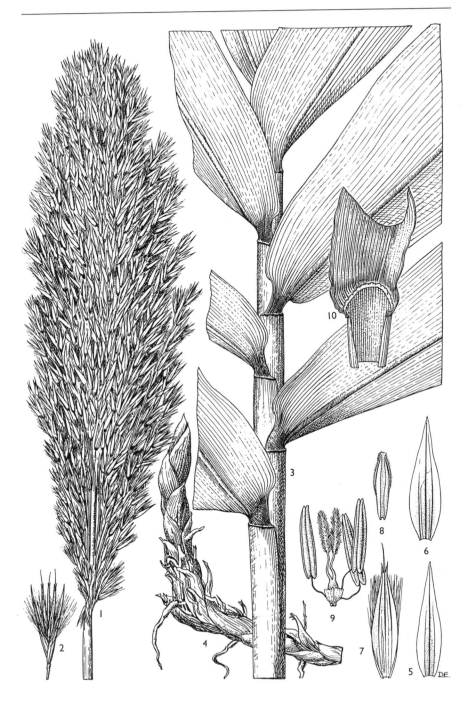

Many grasses in the subfamily Arundinoideae are edible to livestock but relatively few would be those of first choice. Of three species of *Aristida* for example, two are said to be avoided during seed formation, because of sharp points on the lemmas.

3. The small subfamily Centothecoideae (CR 1 tribe, WD place this tribe in Bambusoideae)

Each authority allocates to the single tribe mostly the same genera. Veins, ligule membranous, chlorenchyma often forming a palisade layer below the upper epidermis.

The group is adapted to low-latitude forest shade and its status, whether having a greater affinity to the Bambusoideae or the Arundinoideae, remains controversial.

The basic chromosome number is $x = 12$.

No cereals nor grasses of conspicuous economic importance occur here although *Centotheca lappacea* (Fig. 2.6), for example, is recognised as excellent fodder.

4. Pooideae (CR 10 tribes, WD 2 supertribes, 7 tribes)

This subfamily is uniformly herbaceous and a conspicuously successful colonist of temperate regions.

For the taxonomist, important features include hollow internodes, leaf blades not pseudopetiolate, no cross-veins, ligule an unfringed membrane. The leaf blade in transverse section reveals a uniformly C_3 type anatomy. The basic chromosome number is $x = 7$ although lower numbers are known.

The pooids comprise a large proportion of the grasses and, although differences in both vegetative and reproductive detail do occur, the group is relatively uniform. It takes, to perhaps an extreme, a trend found elsewhere in grasses of the 'all-purpose plant body', extending its range with the assistance of minor variations only. The subfamily is immensely important since it includes wheat, barley, oats, rye and a range of forage and amenity grasses. These include *Agrostis*, *Cynosurus*, *Festuca*, *Lolium* and *Poa*. Additionally *Ampelodesmos*, *Lygeum* and *Stipa* are among the 'esparto' grasses used for paper and rope making. Note, however, the earlier comment on *Stipa*'s taxonomic position. Among ornamental genera are variants of *Alopecurus*, *Briza*, *Deschampsia* and *Festuca* (Fig. 2.7). *Puccinellia* has, in recent years, been used to rehabilitate salinized soil.

Fig. 2.4. *(opposite) Arundo donax* L.
1. inflorescence $\times \frac{2}{3}$; **2.** spikelet $\times$ 2; **3.** part of stem showing leaf bases $\times \frac{2}{3}$; **4.** part of ascending rhizome $\times \frac{2}{3}$; **5.** lower glume $\times$ 3; **6.** upper glume $\times$ 3; **7.** lemma, abaxial view $\times$ 3; **8.** palea $\times$ 3; **9.** flower $\times$ 6; **10.** ligule $\times \frac{2}{3}$.
A. donax occurs commonly throughout the tropics and subtropics especially at lake and river margins. It makes a not very robust cane suitable for screens. Cattle browse its young leaves but it is thereafter only low-quality forage. Variety 'Versicolor' is a striped ornamental.

Fig. 2.5. *Stipa capensis* Thunb.
1. habit × 1; **2.** ligule × 4; **3.** spikelet × 1.75; **4.** lower glume × 3.5; **5.** upper glume × 3.5; **6.** lemma, flattened × 3.5; **7.** palea × 3.5; **8.** fruit × 3.5; **9.** flower × 9; **10.** grain × 3.5.

Fig. 2.6. *Centotheca lappacea* Desv.
1. whole plant × $\frac{1}{2}$; **2.** spikelet × 5; **3.** lower glume; **4.** upper glume; **5.** unarmed lemma; **6.** palea; **7.** flower and palea; **8.** pistil; **9.** armed lemma − all × 6.5.
C. lappacea is a forest grass found in glades and along roadsides. It provides excellent fodder. The spines on the lemma allow distribution on animal fur.

Fig. 2.7. *Festuca killickii* L.

1. habit × $\frac{1}{2}$; **2.** ligule × 2.7; **3a.** spikelet (prior to anthesis) × 2.7; **3b.** spikelet (after anthesis) × 2.7; **4.** floret at anthesis × 5.3; **5.** lower glume × 4.3; **6.** upper glume × 5.3; **7a.** lemma (in plane view) × 5.3; **7b.** lemma (in profile) × 5.3; **8.** palea × 5.3; **9.** androecium and gynoecium of male flower × 8; **10.** androecium and gynoecium of female flower × 8; **11.** lodicules × 8.

F. killickii occurs in subalpine grass veld in South Africa, where it can be locally dominant.

The presence of wheat and barley among the pooids is sufficient reason for this group of grasses to be intensively studied but a feature of particular interest is that among wheat and its relatives (of which barley can be shown to be one) it is difficult to draw taxonomic boundaries. It appears that the group here is in a state of genetic flux – convenient for the breeder, less so for the systematist. For various viewpoints on this situation see Stebbins (1956), Miller (1987), Gupta and Baum (1989) and Chapman (1990).

5. Chloridoideae (CR 5 tribes, WD 4 tribes)
Typically this is a group associated with the drier tropics and subtropics although *Cynodon* occurs in warm temperate areas, and *Spartina* at still higher latitudes on sea coasts.

Technically, important features include culms herbaceous, leaf blades not pseudopetiolate, ligule a fringed membrane or hairs, spikelets laterally compressed, female fertile spikelets usually disarticulating above the glumes. Leaf anatomy in transverse section C_4 and indicating PCK and NAD-ME pathways.

Chromosome basic number is often $x = 10$, sometimes lower.

This subfamily is interesting for several reasons, especially in recent years, because of its increasingly recognised importance in fragile habitats of the drier tropics. Among the various genera are *Plectrachne*, *Symplectrodia* and *Triodia*, the 'spinifex' grasses typical of central Australia.

Among the rangeland grasses are *Astrebla* (Mitchel grasses in Australia), *Buchloë dactyloides* (the buffalo grass of North America), *Chloris gayana* (Rhodes grass of upland central Africa) and *Zoysia* species used in Florida and southern France. *Buchloë* shows separation of the sexes with some plants entirely female, others male and a minority carrying the contrasted male and female inflorescences.

Two cereals, *Eragrostis tef* (t'ef, grown chiefly in Ethiopia) and *Eleusine coracana* (finger millet, a largely African crop), occur here.

A theme running through the Chloridoideae is drought tolerance and this is taken to its extreme in the 'resurrection grasses' – those which revive after dehydration to air-dryness. These include *Eragrostis nindenensis* and *Oropetium capense* in South Africa (Gaff and Ellis, 1974) and in Australia *Eragrostiella bifara*, an undescribed *Sporobolus* species and *Tripogon loliiformis* (Lazarides, 1992). Resurrection grasses occur in the other subfamilies notably in Arundinoideae (*Micraria*) and Pooideae (*Poa*) (Lazarides, *ibid.*).

Many chloridoid grasses show both drought and salt tolerance and some additionally can tolerate relatively high pH.

Again, salt tolerance is not the exclusive preserve of the Chloridoideae among grasses but is none the less conspicuous here. For a detailed list see Chapman (1992b). Among the chloridoids, salt is absorbed from the soil, passes through the plant and is excreted on to the leaves through 'microhairs' that function as salt glands. Such a property has been put to use in *Leptochloa fusca* to remove salt from soil in Pakistan (Islam ul Haq

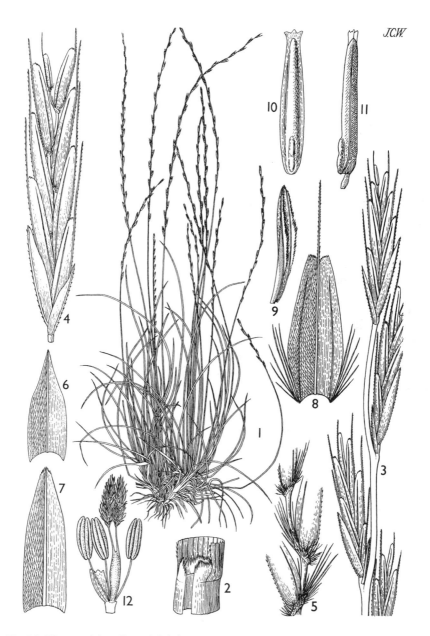

Fig. 2.8. *Tripogon minimus* Roem. & Schult.

1. habit $\times \frac{2}{3}$; **2.** ligule $\times$ 26; **3.** portion of spike $\times$ 6.1; **4.** spikelet $\times$ 8.8; **5.** part of spikelet bases showing floret bases and rachilla $\times$ 12.3; **6.** inferior glume $\times$ 12.3; **7.** superior glume $\times$ 12.3; **8.** lemma $\times$ 17.6; **9.** palea $\times$ 17.6; **10.** caryopsis face view $\times$ 35; **11.** caryopsis side view $\times$ 35; **12.** pistil, stamens and one lodicule $\times$ 26.3

T. minimus occurs in Africa in open and wooded grassland. Like its relative in Australia, *T. loliiformis*, *T. minimus* is an example of a 'resurrection' grass.

and Khan, 1971) and in India (Rana and Parkash, 1980). The method, to be successful, requires harvesting and removal of the salt-crusted foliage. A concluding instance of a salt-tolerant genus is *Spartina*. The colonisation of *Spartina* around the British coastline is relatively well documented (Huskins, 1931), and Huskins' work provided the first cytogenetic evidence of evolution of new species in nature within historical times. *S. maritima* is indigenous. The introduction of *S. alterniflora* in the early nineteenth century appears to have triggered the following events.

$$
\begin{array}{ccc}
S.\ maritima & \times & S.\ alterniflora \\
2n=60 & & 2n=62
\end{array}
$$

Before 1870

$$\downarrow$$

S. townsendii   (empty pollen
2n = 61   grains)

Before 1890

$$\downarrow$$

S. anglica   (full pollen
2n = 122   grains)

S. alterniflora, indigenous to North America, has been used to stabilise coastlines. Broome *et al.* (1986) point to the importance of tidal marshes as habitats for wildlife, in absorbing excess nutrients from sewage pollution, in dissipating wave energy and in reducing shoreline erosion. Deliberate planting of *S. alterniflora* as a cheaper alternative than bulkheads, groins and revetments makes an interesting parallel to the use of vetiver grass instead of expensive stone terracing discussed in Chapter 10. Broome *et al.* (*ibid.*) describe the successful stabilisation of a marsh habitat at Pine Knoll shores on the coast of North Carolina. As these authors point out, such a technique is relevant to the problems created by gradually rising sea level.

6. Panicoideae (CR 7 tribes, WD 2 supertribes, 6 tribes)

This subfamily is herbaceous and centred in the tropics and subtropics with occasional outliers in the temperate regions.

Technically important features include a ligule of various sorts, spikelets single or paired, each usually two flowered. The two florets are often dimorphic, the upper one tending to be hermaphrodite, the lower one male or barren. Photosynthetic pathways are diverse and include C_3 and C_4 (including PCK, NAD-ME and NADP-ME). The basic chromosome number is typically $x=9$ but $x=16$, 10 and 7 occur. Panicoid and andropogonoid examples are given in Figures 2.9 and 2.10.

A noteworthy feature here is that two tribes recognised by Watson and Dallwitz, the Steyermarkochloeae and the Eriachneae placed by them in Arundinoideae, are considered by Clayton and Renvoize to belong to the Panicoideae. Each of these systems allot the same genera to each tribe, two in each case, suggesting that the tribes themselves internally cohere.

Within the Panicoideae, the interpretation of Watson and Dallwitz (1992) of two supertribes, Panicodae and Andropogonodae, emphasises trends to which Clayton and Renvoize (1986, 1992) give rather less importance. None the less, for each system the andropogonoids are viewed as distinctive and incorporate substantially the same genera. A feature of this group is that type of C_4 photosynthesis described as NADP-ME (XyMS-) and where, for chromosome number, $x = 10$ or 12. Andropogonoid grasses have their spikelets typically in pairs (1 sessile and 1 pedicellate), an arrangement illustrated in detail in Chapter 4 and which underlies part of the interest of maize evolution discussed in Chapter 12.

Among forage grasses are *Brachyaria decumbens*, *Cenchrus ciliaris* (buffel grass), *C. setigerus* (birdwood grass), *Panicum maximum* (Guinea grass), *Paspalum notatum* (Bahia grass), *Pennisetum purpureum* (elephant, Napier grass) and *Saccharum sinense* (uba cane). *Hyparrhenia hirta* (Fig. 2.10), although of moderate grazing value, is widely distributed on poorer soils. It can be used for thatching.

Cereals include *Coix lachryma-jobi* (adlay, Job's tears), *Digitaria exilis*, *D. iburua* (fonios), *Echinochloa crus-galli* (Japanese millet), *Panicum miliaceum* (proso millet), *Pennisetum glaucum* (bulrush, pearl millet) and *Sorghum* and *Zea* (maize, Indian corn).

Other useful grasses include *Axonopus compressus* and *Stenotaphrum secundatum*, both used for tropical lawns and, of major importance, *Saccharum officinarum* (sugar cane).

Secondary Metabolites

Across the Poaceae the synthesis of secondary metabolites such as might be useful in medicine is uncommon. The following are some exceptions. *Agropyron repens* (couch grass) rhizomes yield a diuretic of which Boesel and Schilcher (1989) report 61 constituents. *Cynodon dactylon* rhizomes also yield a diuretic. Some grasses, notably *Lolium perenne* (perennial ryegrass) and to a lesser extent *Anthoxanthum odoratum* (sweet vernal grass), *Cynodon dactylon*, *Dactylis glomerata* (cocksfoot, orchard grass), *Holcus lanatus* (Yorkshire fog, velvet grass) and *Phleum pratense* (Timothy), generate pollen allergens (glycoproteins) that induce hayfever (Knox and Singh, 1990). Although in *L. perenne* the allergen can comprise 5% of the soluble protein in the pollen grain, apart from its incidental allergenic property it has no obvious function.

In etiolated shoots of sorghum a cyanic glucoside 'dhurrin' is sometimes produced which, on hydrolysis, yields prussic acid and the risk of cattle poisoning. A remedy has been to select for low dhurrin values (Doggett, 1988).

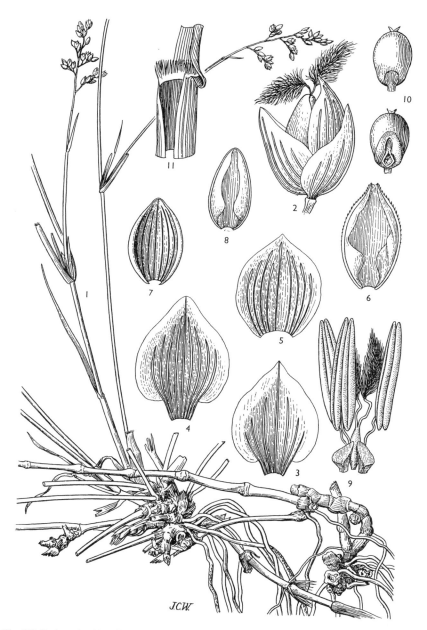

Fig. 2.9 *Panicum turgidum* Forssk.

1. habit × $\frac{2}{3}$; **2.** spikelet × 7.1; **3.** lower glume × 8.9; **4.** upper glume × 8.9; **5.** lower lemma × 8.9;
6. its palea × 8.9; **7.** upper lemma × 8.9; **8.** its palea × 8.9; **9.** flower × 13.3; **10.** grain × 8.9;
11. ligule × 10.6.

P. turgidum, found in the Saharo-Sindian region, forms dense bushes up to 1 m tall. Its roots have fine hairs
that accumulate sand particles giving a felty appearance. The robust foliage, which is able to withstand
severe desiccation, provides camel fodder and the grains are sometimes collected for food.

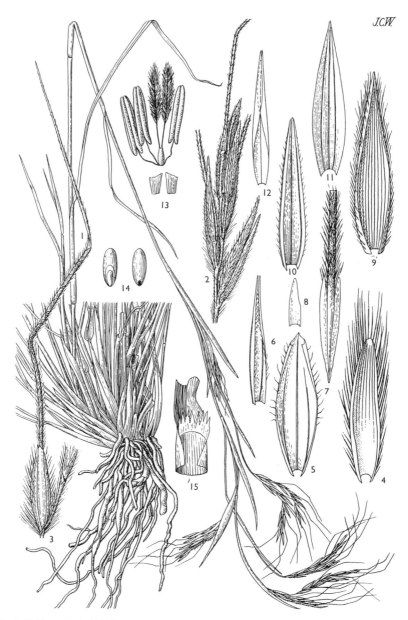

Fig. 2.10. *Hyparrhenia hirta* Fourn.

1. Habit × > $\frac{2}{3}$; **2.** portion of the inflorescence showing the lower homogamous spikelets × 4.8; **3.** pair of spikelets × 4.8; **4.** lower glume × 7; **5.** upper glume × 7; **6.** lower lemma × 7; **7.** upper lemma with base of awn × 7; **8.** its palea × 7; **9.** lower glume × 7; **10.** upper glume × 7; **11.** lemma × 7; **12.** palea × 7; **13.** flower × 7; **14.** grain × 2.5; **15.** ligule × 3.5.

H. hirta is a dense tufted perennial surviving well on waste ground and is likely to replace less vigorous competitors.

Among other secondary products are coumarin, an aromatic hydroxy-acid derived for perfumery from *Anthoxanthum odoratum*. Other sources of perfume derive primarily from the genus *Cymbopogon*. The following are the most important: *C. nardus, winterianus* (citronella oil), *citratus, flexuosus* (lemon grass oil), *martinii* (palmarosa oil), *nardus* var. *confertiflorus × jwarancusa* (jamrosa oil) and *martinii* var. *sofia* (gingergrass oil). Additionally, there remain other fragrant species currently of minor consequence but perhaps likely to increase in importance. These include *C. densiflorus, distans, microstachys, jwarancusa* (as a pure species), *proximus, sennarensis, coloratus, afronardus, shoenanthus* (poisonous), *georugii, polyheuros, giganteus* and *khasianus*. For a full account see Boelens (1994).

The advent of gas–liquid chromatography and mass spectroscopy have considerably increased our knowledge of essential oil composition among these grass oils. Up to 90 ingredients have been recognised for a given oil. Within a species varieties differ and the environment in which they grow can also exert influence on oil composition. New collections are still being made, for example in the Himalayas, and the effects of deliberate hybridisation remain largely unexplored. *Vetiveria zizanioides* yields vetiver oil from its roots. Alternatively, such roots spread over a stone or tile floor and sprinkled with water provide a fragrant surface upon which to walk. Among the components of vetiver oil are α and β-vetivones, vetivenyl vetivenate, vetivenic acid, palmitic acid, benzoic acid, vetivene, khusol, khusitol and khusinol.

A further group of chemicals is generated among grasses by infection with ergot. The fungus *Claviceps purpurea* creates ergot in rye whereby grains become black, enlarged and misshapen. Alkaloids form which when extracted and applied in pregnancy can relax the uterus. Traditionally, 'St Anthony's fire', an hallucinatory condition, is ascribed to people eating bread made from ergotty rye. Other ergots generate different alkaloids and *Cenchrus echinatus* infected with *Balansia obtecta* will yield ergobalansine (Powell *et al.*, 1990). Some endophytic fungi are now known to infect grasses though with few visible symptoms. They none the less generate secondary metabolites increasingly recognised as important.

Fungal Endophytes

During the 1970s it became apparent that some grasses were detrimental to livestock, the effect being due to fungi living within the grass tissues. The principal interests here are to identify the pattern of infection and the consequences for grazing herbivores. Integral to this is the notion that such fungi imparting reduced palatability contribute to a grass survival strategy along with ready establishment, spreading or tiller growth, resistance to desiccation, concealed meristems and accumulation of silica. For a discussion see Clay (1984), Chaplick and Clay (1988), and Vicari and Bazely (1993).

This group of fungi includes the ascomycete genera *Atkinsonella, Balansia, Epichloë* and *Myriogenospora*, all of which demonstrate a sexual stage. Asexual, closely related endophytes, are grouped under *Acremonium* section *Albo-lanosa*. *Atkinsonella* occurs on *Danthonia* and *Stipa*, both placed by Watson and Dallwitz (1992) in *Arundinoideae*. *Stipa*, however, is regarded by Clayton and Renvoize (1992) as belonging to Pooideae. Just how useful fungus host ranges are as a measure of relationship is explored subsequently using rusts and smuts. In passing, for example, it is worth noting that *Epichloë* and its close relative *Acremonium* infect pooids. They do however also infect *Brachyeltrum*, normally included with bamboos. Beyond this, though, at present nothing appears to be known about fungal endophytes in the Bambusoideae.

The infection patterns among fungal endophytes vary. *Balansia* infection may drive a plant asexually by inhibiting its sexual reproduction. Among several grasses, for example *Andropogon, Festuca* and *Poa*, endophytes can induce prolifery (vivipary) although not every viviparous grass is necessarily infected. At a less conspicuous level there appears to be a link between infection and a change from chasmogamy to cleistogamy though few sufficiently detailed case histories are available. An exception is a study by Clay (1984) of *Danthonia spicata* infected with *Atkinsonella hypoxylon*. For infected plants, tiller number more than doubles and the inflorescence, normally chasmogamous, becomes entirely cleistogamous. Since cleistogamous flowers can set seed (which is both viable and infected) the process is repeated in subsequent generations. As Clay (*ibid.*) points out, infection thus imposes recurrent inbreeding on an otherwise mostly outbreeding population. The endophyte is inherited in a way which mimics that for a maternal gene.

In endophytic infection is the relationship 'parasitic' or 'mutualistic'? In the latter case 'fitness' in the genetic sense increases. The conclusion need not be the same for each grass/fungus association.

It is clear that photosynthetic performance can be altered and auxin has been found in pure cultures of *Balansia epichloë* (Porter *et al.*, 1985). Of more immediate concern are the poisonous properties of infected grasses for livestock. Bacon *et al.* (1986) provide evidence for *Balansia* of 20 alkaloids and record its infection in *Agrostis* (1), *Andropogon* (2), *Calamagrostis* (3), *Chasmanthum* (1), *Chloris* (2–3), *Ctenium* (1), *Eragrostis* (6), *Gymnopogon* (1), *Oryzopsis* (1), *Panicum* (3), *Sporobolus* (3), *Thraysia* (1) and *Triodia* (1) (species numbers of grasses in brackets). Where unthriftiness in cattle can be ascribed to endophytic fungi, the production of clean pasture from seed is complicated by wild grasses nearby providing a reservoir of infection.

An example quoted by Bacon *et al.* (*ibid.*) is as follows. In New Zealand infected ryegrass pastures resist the depredation of the Argentine stem weevil *Listronotus bonariensis* but are poisonous to cattle. Uninfected ryegrass is so susceptible to the depredations of the weevil that it does not provide adequate grazing for cattle.

This consideration of fungal endophytes serves to emphasise that grass diversity can be not only genetic or environmental in the customary sense but also derives from these less straightforward causes.

Diversity and Relationship

When considering the diversity of grasses, questions inevitably surface about relationship. Can one, for example, put the five (or six) subfamilies of grasses into any convincing kind of evolutionary order? If trimerous flowers are considered more primitive then the Bambusoideae would be, presumably, representative of the oldest surviving grasses. Conversely, Clayton and Renvoize (1992) have attributed primitive, relict status to the Arundinoideae. Given deforestation in the tropics at present, the more weedy aggressive members of Panicoideae are advancing in many habitats and create the impression of widespread success.

In fact, the relationships among the subfamilies do not admit of easy arrangement. One way of indicating the complexity of the situation is to offer two of 'nature's own assessments'. Certain fungi, notably rusts and smuts (Basidiomycetae, Order Ustilaginales, Families Pucciniaceae and Ustilaginaceae) infect grasses but do so selectively. The smut genera *Urocystis* and *Neovassia*, for example, infect in the former case only Pooideae and Arundinoideae and in the latter case only Arundinoideae. Based on Watson (1990) Fig. 2.11 indicates host preferences of five rust and ten smut genera among the five major grass subfamilies.

One might assume that, since the rust *Puccinia* and the smuts *Tilletia* and *Ustilago* affect representatives of all subfamilies, they have accompanied the Poaceae from its early inception and precede its subfamilial divisions. Conversely, *Neovassia* which is peculiar to the Arundinoideae could be seen as post-dating the evolution of that subfamily. This is the more speculative approach. The less speculative approach is to recognise that, at the present time, Chloridoideae and Panicoideae have more of these pathogens in common than does either of them with any other subfamily and they are perhaps therefore more closely related. Conversely, on this basis the Pooideae appear relatively isolated. None the less, puzzling anomalies remain. *Urocystis* will infect Pooideae and Arundinoideae, while *Entyloma* will infect Pooideae and all the other subfamilies except Arundinoideae. Perhaps, therefore, the situation in regard to rusts and smuts provides us

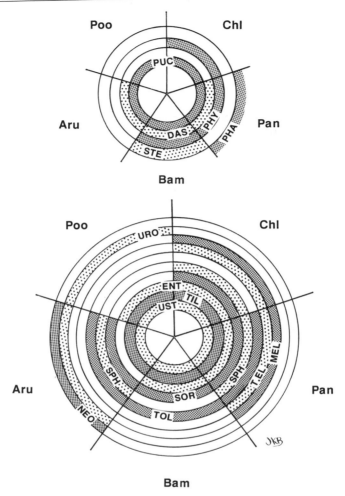

Fig. 2.11. Host preferences among the five grass subfamilies of five rust genera (DAS *Dasturiella*, PHA *Phakopsora*, PHY *Physopella*, PUC *Puccinia*, STE *Stereostratum*) and ten smut genera (ENT *Entyloma*, MEL *Melanotenium*, NEO *Neovassia*, SOR *Sorosporium*, SPH *Sphacelotheca*, TIL *Tilletia*, TOL *Tolyposporium*, T.EL *Tolyposporella*, URO *Urocystis*, UST *Ustilago*).

with a cautionary tale against heady generalisations about subfamily relationships.

Below subfamily level it is possible to find at tribal, generic or species level instances of small and large variability. Some tribes are small and restricted in distribution, their few species being confined to highly specialised situations. The tribe Centotheceae with its 11 genera is an obvious instance. At the other extreme, in the tribe Triticeae in the Pooideae and, within the confines of one species, *Zea mays*, so great is the variation available that these situations have been described as instances of 'genetic turbulence' (Chapman, 1990).

Although the accompanying diagrams to this chapter illustrate some of the intricacies of panicle, spikelet and floral detail, the aim here is to provide a broad awareness of the various subfamilies. In addition to the structural detail required to provide a basis for systematic arrangement, variation in other directions also occurs. Merely from structural evidence it would be difficult to explain why one species was weedy and widely dispersed and another far less so. There exists a diversity at reproductive level that hardly follows patterns confined within subfamilies. Such diversity is important and is treated separately in Chapter 9. A prerequisite is to understand something of grass development from seed to maturity.

The Assumption of Form 3

An individual grass seed is normally shed with its ovary as a single unit, the 'caryopsis' and this to a greater or lesser extent is associated with 'chaff', the dried remains of glumes, lemma and palea, stamen filaments, lodicules and stigmas. Of these, the lemma can be of special interest being extended sometimes into an awn and even occasionally into a plume functioning as a parachute. Various assumptions have been made about the awn (since it is both spiral and hygroscopic), for example that when hydrated it can move the caryopsis along the ground and drill it into the soil. These latter are examined in some detail later in Chapter 5.

Embryo Shape and Size

At the conclusion of embryogenesis and before germination, the embryo is closely associated with the endosperm food reserve. The shape, size, orientation and relative proportion of embryo to endosperm tends to be characteristic of particular grass groups (Reeder, 1957, 1962: Mladá, 1974). Figure 3.1 shows a generalised grass embryo, Fig. 3.2 a newly germinated maize seed and Fig. 3.3 a later stage for a germinated maize seedling. The legend interrelates the three figures.

Germination

In general terms, the process is a familiar one, hydration of the seed leading to the emergence of root and shoot. However, germination of grasses raises questions of more specialised interest.

The coleorhiza is problematic. It at first sheathes the emerging radicle but is itself able to absorb water and mineral salts. Morphologically its

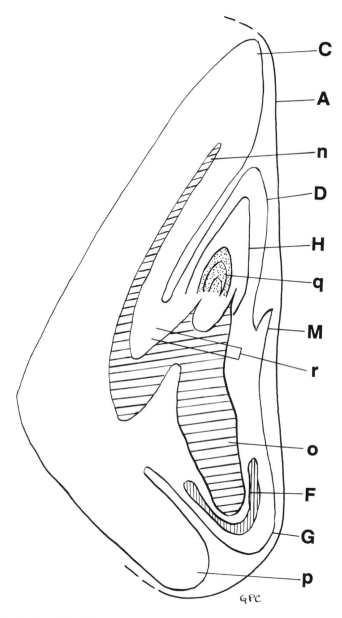

Fig. 3.1. Vertical sectional through a generalised grass embryo before germination. The region to the left of the scutellum is occupied by endosperm.

Figs 3.1, 3.2 and 3.3 use upper-case letters to indicate features eventually visible externally. Lower-case letters indicate features remaining concealed.

A–pericarp; B–remnant of style; C–scutellum; D–coleoptile; E–ruptured pericarp; F–radicle; G–coleorhiza; H–first true leaf; I–primary root branch; J–emerging adventitious roots; K–primary root hairs; L–adventitious root hairs; M–epiblast; n–scutellum vascular strand; o–root vasculation; p–hypopeltate appendix (the region between p and G is the scutellum cleft); q–plumule; r–mesocotyl.

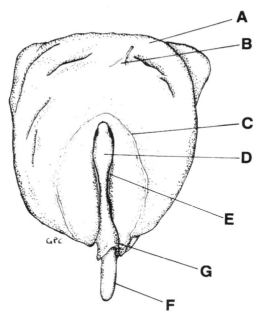

Fig. 3.2. A newly germinated maize seed. (See Fig. 3.1 for key to labels.)

status is uncertain, being regarded either as an extension of the hypocotyl (Velenovsky, 1907), a modification of the primary root (Brown, 1960), or a 'new' organ, a proposition normally accepted only reluctantly by botanists.

The coleoptile, too, has also provoked alternative interpretations. Celakovsky (1897) proposed, ingeniously, that if the scutellum corresponds to a leaf blade, then the coleoptile is a modified ligule and the epiblast is homologous with the leaf auricles. Arber (1934) took a different view arguing that the coleoptile is a leaf with two (outer) vascular bundles the middle one having been diverted to the scutellum (which, of course, then questions the status of that structure as a cotyledon).

The scutellum, nowadays commonly regarded as a cotyledon, was in the past considered an excrescence of the hypocotyl, a blindly ending stalk. If this were so and one is committed to grasses being monocotyledons then the epiblast is a cotyledon candidate unless of course there is no epiblast as is the case in a number of grasses.

Since the scutellum, coleoptile and epiblast are in close proximity they may have some kind of relationship. In addition to ideas already mentioned others have included the following:

> The scutellum, coleoptile and epiblast are not homologous with leaf structures but are specific to the grass embryo.

Brown (1959)

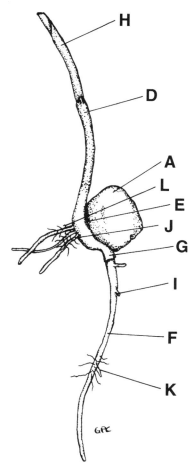

Fig. 3.3. A later maize seedling showing emergence of first true leaf. No epiblast is evident. (See Fig. 3.1 for key to labels.)

The epiblast (unique to grasses) is a rudimentary second cotyledon and grasses are therefore 'heterocotyledonous'.

Roth (1959)

The epiblast is a ligule.

Boyd (1931)

The epiblast (together with the coleorhiza) is an 'excrescence' of cotyledon or of the embryonic stalk.

Sargant and Arber (1915)

Scutellum and coleoptile correspond to cotyledon and ligule with the epiblast as a non-vascular excrescence.

Arber (1934)

The problem of homology here is unresolved but the present writer offers his tentative support for Celakovsky (1897) in the absence of anything more convincing. For a detailed discussion see Mladá (1974).

A further problem is with the hypopeltate appendix or scutellum cleft. It could be regarded as a consequence of enlarging the absorptive area of the scutellum. Since it is missing in two groups, the oryzoid and pooid grasses, and assuming the bambusoid situation, where it is present, to be primitive, one needs to stretch ingenuity to explain how it was abandoned. Alternatively, here as elsewhere one might allow that incidental, as well as useful, features survive the effects of natural selection.

At least as perplexing as any of the foregoing is the mesocotyl distinguished first by van Tieghem (1897). It is located (when it occurs) *above* the point of cotyledon insertion on the main axis. Why not then simply regard it as an epicotyl? The reason is that its vascular arrangement resembles a hypocotyl (i.e. it is more root-like) as might be expected if it occurred *below* the point of cotyledon insertion. 'Mesocotyl' therefore is a term that indicates an ambiguity about its status. It does however help to separate groups of grasses by its presence or absence.

Relationships among the subfamilies

In the previous chapter the relationships among various rusts and smuts in their infectivity toward the grass subfamilies were set out. Much the same equivocal situation emerges from a consideration of first leaf roll, epiblast, scutellum cleft and mesocotyl representation (Fig. 3.4). Each group represented differs from the bamboos in one or more respects but no one character seems inextricably linked with any other. A rolled first leaf can occur with or without an epiblast and a scutellum cleft can appear with or without a mesocotyl. Conversely, within the subgroups represented the characters present are relatively consistent. There are exceptions. Among the Pooideae, for example, species exist without an epiblast in *Arctophila*, *Agropyron*, *Brachypodium*, *Bromus*, *Dupontia*, *Elymus*, *Hystrix*, *Leymus*, *Secale* and *Taeniatherum*. Since efficient functioning of the embryo is crucial for the life cycle, we see here associations of characters deeply embedded in the genetic history of the subfamilies and presumably of some significance for their survival though in what way is far from clear.

Having digressed, we return to the process of germination. As hydration proceeds, resources are mobilised from the endosperm through the scutellum and root and shoot extension begin protected by the coleorhiza

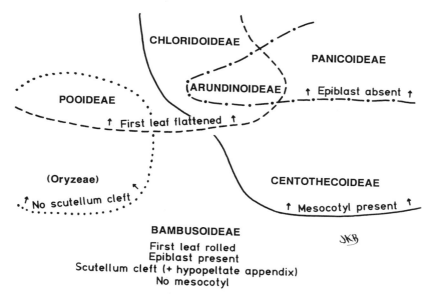

Fig. 3.4. Distribution of seedling characters in the Poaceae.

and coleoptile respectively. Growth of the primary root system is normally brief and it is progressively replaced well after germination by adventitious roots arising from the lower stem nodes. The coleoptile opens and through it protrudes eventually, either rolled or flattened, the first true leaf (Fig. 3.3).

Emergence and Establishment

The largest known fruit of grass, some 7–12 cm long, is that belonging to the bamboo *Melocanna baccifera,* and after germination its seed can establish a primary shoot in excess of 5 m. There is, however, no clear-cut relationship between seed size and the eventual dimensions of the stem. Seeds of some *Dendrocalamus* species and bread wheat (*Triticum aestivum*) are not greatly dissimilar in size and *Zea mays* exceeds both. In final size *Dendrocalamus* species (bamboos) form woody specimens vastly bigger and more long-lived than either of the two cereals. There is an interesting complication. Since monocotyledons lack a secondary thickening system derived from persistent and rejuvenated meristem how can the massive stem of a bamboo derive from so small an apical meristem?

Seed germination, a relatively rare event in some bamboos, results in the emergence of a primary root and shoot, the latter as slender as that of many other grasses. The primary root is short-lived but so too is the shoot. Before its demise, there is established from its base rhizomatous growth from which, successively, larger and larger aerial stems arise. In *Dendrocalamus* the plumule emerges to give a slender shoot with sheathing scale-like leaves and appearing as only the most diminutive version of the later massive culms that will eventually arise. From its base rhizomes arise that bend downwards into the soil before turning upwards to produce a subsidiary though larger culm. The process repeated by successive rhizomes results in progressively larger diameter and taller culms being originated from deeper in the soil until the situation stabilises at the final culm size.

A feature of special interest is the 'neck'. A bud arising from a rhizome elaborates a series of closely packed nodes of which each successive one is of greater diameter and creates a situation whereby the eventual culm diameter exceeds substantially the rhizome from which it arose. Typically, the bamboo provides a situation where *both* primary root and primary shoot are of short-lived consequence, their places being taken by adventitious roots that arise from an enlarging rhizome/culm complex. The developmental detail that is associated with progressively larger aerial stems seemingly awaits close histological investigation.

Suppose that the situation regarding the primary root is unchanged but the primary shoot having emerged continues to grow and any secondary axillary buds arising from its base remain precisely that, namely of secondary consequence. If the primary shoot not only retains its primacy but, after a relatively short interval flowers and fruits, we could interpret this as 'neoteny' whereby what was a 'bamboo' life-style has been dramatically telescoped as is the case in other grasses. The advantages seem considerable. Not only does the rhizome, or at least tillering ability, remain, but the adoption of a less 'cumbrous' plant body allows more 'seasonal' environments to be exploited and, more importantly, the generation interval is considerably shortened permitting greater responsiveness to natural selection. It is not, however, mandatory to read evolution this way. Since the woody stems of bamboos have no obvious counterparts in other plant families, they might be regarded as a derived condition from more herbaceous ancestors and a relatively late response to forest dominance by more conventionally woody species.

The Developing Plant Body

Key features of germination are the emergence of the primary root through the coleorhiza and emergence of the first leaf through the coleoptile. Thereafter the primary root can branch, a phase normally of brief duration. The stem apex shows three detectable phases, namely differention of leaf sheath primordia, the establishment of a detectable nodal pattern and the development of bud primordia in the sheath axils. At the base of each node meristematic activity persists longer. It is from these latter regions that adventitious roots (those of stem origin) arise and, sooner or later, replace the primary root system as effective absorbing organs of water and mineral salts (Fig. 3.5).

Because of the importance of grasses for pasture, events surrounding the elaboration of the leaf have been minutely detailed. The generalised sequence of events is as follows. Initially, the leaf primordium is entirely meristematic and develops as a flap of tissue over-arching and encircling the stem apex. Meristematic activity is then progressively confined to an intercalary region that elaborates on its inner (adaxial) face the ligule and, more abaxially, the leaf lamina, a point to which we return later in discussing the lemma.

If the meristem at the base leaf sheath and even the intercalary meristem adjoining the ligule remain protected by the soil surface or too low to be accessible to grazing animals, removal of the leaf tip does not preclude further leaf growth. Contrary to an assertion sometimes made,

Fig. 3.5. Adventitious roots arising from the lower nodes of a maize culm.

removal of the upper leaf parts do not *stimulate* further growth. Apparent regrowth is due to the emergence of erstwhile concealed basal parts (Begg and Wright, 1962).

The shoot apex differentiates in its wake a stem with leaves at alternate nodes opposite each other. Every *other* node therefore has a leaf on the *same* side of the shoot. The region between the nodes is, typically, hollow in pooid grasses and solid in panicoids. Bamboos, depending on species, can have either hollow or solid internodes and both can occur in the same plant (e.g. *Chusquea*).

Branches and branch complements

Although grasses can produce occasional branches from the upper nodes, as for example with the highly compacted lateral branch bearing the maize ear, most often what occurs is growth from the stem base. This branching can take two forms. If the bud develops within the arc of leaf tissue it will tend to grow vertically and produce a tiller. Such growth is 'intravaginal' – literally 'within the pouch'. Alternatively, the developing bud may break through the leaf base and form a stolon or rhizome and is said to be 'extra vaginal' or 'beyond the pouch'.

Bamboos here as elsewhere provide an interesting exception. In the upper nodes a cluster of branchlets can arise known as a 'branch complement' (Fig. 3.6). In *Phyllostachys* the complex consists of two branches one larger and one smaller. More commonly there are several with perhaps one dominant as occurs in *Atractantha*. A feature associated with branching is the 'prophyllum', but this is more conveniently dealt with later in Chapter 4.

Bamboo leaf variation

Typically a grass, in the elaborating shoot, sets off a series of similar leaves, there being little variation in size or shape with perhaps the exception of the 'flag' leaf from within which the inflorescence eventually emerges. Even if the plant remains vegetative, producing a series of tillers or stolons, variations in leaf shape and size are minor.

By contrast, again there is considerable diversity among the bamboos in matters of detail. One situation is where on successive nodes up a bamboo culm there is a progressive elaboration of leaf shape. Those below consist only of a leaf sheath and may be barely photosynthetic or reduced to scales. Higher up the lamina forms a greater proportion of the leaf and is an effective photosynthetic organ. In *Sasa*, by contrast, the changeover from scale to photosynthetic leaf is relatively clear cut (Fig. 3.7). Perhaps the

Fig. 3.6. Branch complement in *Bambusa rigida* seen against the background of a partly abscised scale (non-photosynthetic) leaf in the axil of which it developed.

strangest of all situations among the grasses occurs in the South American bamboo *Glaziophyton mirabile*. Initially it produces a stem consisting largely of one node reminiscent of a rush even in regard to its pithy septate interior. After burning, leafy tillers, recognisably more bambusoid, develop. It is the latter kind of culm that bears the inflorescence. It is as yet unclear whether the rush-like growth habit is a juvenile phase re-established after each germination. *G. mirabile* is known only in two isolated populations for which field observations are minimal. For an account of this remarkable grass, see McClure (1973).

For many bamboo species after establishment there follows that process, seemingly interminable, of one culm after another. And then

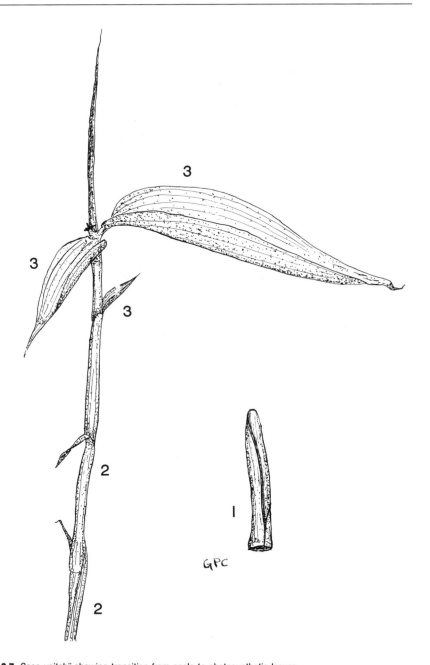

Fig. 3.7. *Sasa veitchii* showing transition from scale to photosynthetic leaves.
1. scale leaf near culm base consisting of sheath only × 1; **2.** scale leaf with sheath and rudimentary lamina; **3.** photosynthetic leaves × $\frac{4}{5}$.

suddenly things change and a flowering shoot is produced that may consume the plant's energies completely. McClure (1966) draws a distinction between vegetative maturity when full-sized culms are produced and sexual maturity marked by panicle emergence and, subsequently, seed production.

Agronomy

For cereals and grasses raised commercially a key factor has been to intervene, preferably early, in the life of the plant with a view to influencing productivity.

The interrelations of root and shoot and the stages through which the shoot can pass depending on nutrient supply and the influence of various environmental factors provide the basis of grass agronomy and finds its expression in improved methods of management for cereal and pasture production. For detailed treatment of these aspects many texts are available including that of Milthorpe and Ivins (1966). Other useful sources are Jones and Lazenby (1958), Cooper and Morris (1983) and Gallagher (1984), all produced from a largely temperate perspective. Texts dealing with cereal and forage production at tropical and subtropical latitudes include Crowder and Chheda (1982). Each of the major cereal crops – maize, rice and sorghum – have stimulated the production of many volumes that are too numerous to detail here. A style of crop literature that perhaps reached its apogee around 1980–1990 was, the so-called 'blueprint' approach to cereal growing, where each step in development was related to the precise application of various chemicals for weed, pathogen and pest control. Such an approach is now considerably modified due to a combination of environmental concern and the availability of cereals in sufficient quantities, at least in Western Europe and North America. The terms 'de-intensification' and 'extensification' describe the present approach to crop management but clearly they imply a transitional situation. As regards the use of *Lolium perenne* in pasture production, it owed its widespread use partly to its nitrogen fertiliser responsiveness. With diminished nitrogen application the performance of perennial ryegrass approaches that of other grasses with, on environmental grounds, the added advantage that a lower nutrient status encourages species diversification and consequently more ecologically interesting grassland with a significant addition of colourful dicotyledons. For a detailed study of the vegetation dynamics associated with grassland extensification in the context of a commercial farm see, for example, Mitchley (1994).

Shoot Anatomy

The seed described earlier eventually elaborates the plumule shoot that may or may not be replaced, or accompanied by secondary shoots. The term 'shoot' is chosen advisedly to make the point that stem and leaf, although examined separately for convenience, nonetheless comprise parts that function jointly. To underline this the terms 'phyton' (quoted by Arber, 1934) and 'phytomer' (Clark and Fisher, 1987), each referring to a leaf, the subadjacent internode, the node and its lateral bud with adventitious roots, if present, have been coined. Although a useful starting point, it is hardly adequate since a leaf interacts not with one node complex but several and, although extremely complex in practice, it is more realistic to view the shoot as a whole.

Grass stems tend in 'pooids' to be hollow and in 'panicoids' to be solid – the former providing the original drinking straws commercially available until about 1950 in the United Kingdom, the latter the solid stems of sugar cane. The distinction is not invariably clear cut and a solid stem mutant in wheat (pooid) has been used to control *Cephas cinctus* whose larval stage incubates in the internode (Wallace *et al.*, 1974). In hollow internodes the vascularisation is confined to the cylinder wall and in solid stems 'scattered' partly through the central pith. 'Scattered' is a misleading term here since the bundle arrangement is both orderly and complex with intricate anastomoses (fusions) at each node.

Vascular Establishment, Pathway and Variation

At the shoot apex, cell divisions establish a trail of differentiating nodes and internodes such that, viewed from one side, alternate nodes originate leaves each subtending an axillary bud. On the opposite side the remaining nodes develop similarly as is characteristic of a 'distichous' arrangement. Stem diameter is fixed although in the familiar bamboo garden canes the higher later-formed nodes and internodes are obviously more slender. The essential concept is that any node or internode once differentiated has a fixed diameter and no prospect through secondary thickening, unlike dicotyledons, of any increase.

From the work of Hitch and Sharman (1971) based on a group of pooid (hollow-stemmed) grasses (*Agropyron junceiforme*, *A. repens*, *Avena sativa*, *Bromus carinatus*, *Cynosurus cristatus*, *Dactylis glomerata*, *Lolium perenne*, *Poa annua*, *Secale cereale* and *Triticum aestivum*) the following pattern emerges. It is accomplished prior to internode expansion. Figure 3.8 indicates the two orthostichies, one each side of the stem in simplified form since only the median vascular traces are shown.

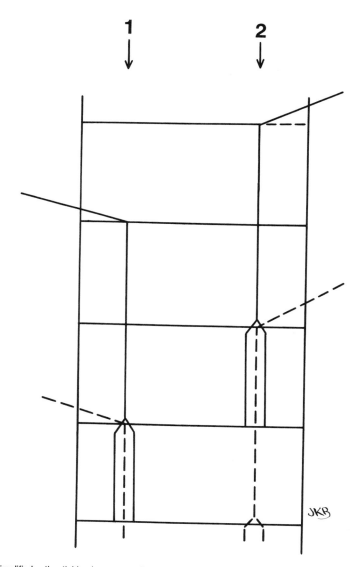

Fig. 3.8. Simplified orthostichies in a grass shoot.

1. As each leaf initial matures, so within it a median vascular trace forms (as do laterals referred to later).

2. This trace is elaborated progressively toward the stem into and down the stem and eventually integrated with the stem vascular system (since the direction of elaboration is downwards away from the apex it is 'basipetalous').

3. As elaboration proceeds down the stem it passes through one node, to bifurcate just above the trace from an older leaf entering at the next node down.

4. In summary, therefore, the trace enters at its 'own' (first) node, passes through one node (second) unaffected, bifurcates just above the next node down (third), passes through it and joins the nodal plexus one more node down (fourth). The nodal plexus is the anastomosis there of vascular strands.

5. What holds true for the median vascular leaf trace is also true for the lateral leaf traces either side of the median, of which there are many.

6. Turning now to the other half of the stem, a similar sequence of events occurs except that it is one node 'out of step'.

7. The concluding step is the elaboration of a vascular connection from the nodal plexus to the axillary bud, thus linking the bud not primarily to its subtending leaf but to the one above it on the opposite side of the stem. If the bud were to develop into a subsidiary shoot it would replicate the events just described.

Upon this general pattern to what extent are there taxonomic differences? Hitch and Sharman (*ibid.*) report seemingly rather minor differences, in the same shoot for example and among the species represented, but claim the generalised pattern described here is inclusive. By contrast *Zea mays* is a solid-stemmed grass. Esau (1964) recognised there that a leaf trace on one side of the stem can, between nodes, cross, on its descent, to the other side.

Some information of this kind is available for bamboos (Grosser and Liese, 1971). It is worth remarking that *Apoclada*, *Aulonema*, *Chusquea*, *Myriocladus* and *Olmeca*, for example, have both hollow and solid-stem variants and might provide suitable test materials.

A further point is that the array of primordia at a node forming the branch complement implies a greater vascular complexity there.

Nature of the Vascular Strand

Typically, in transverse section the vascular strand is as follows. Internal to the phloem with its sieve tubes and companion cells, are two large metaxylem vessels separated by smaller ones and, on the inner side, protoxylem to some extent disrupted. In bamboo stems, the ground tissue becomes substantially lignified, hence its toughness and durability. For a detailed study of contrasting bamboo vascular strands see Grosser and Liese.

The strand, in leaves, may be surrounded by a photosynthetic sheath, a subject considered in detail in Chapter 8 on photosynthesis.

Root Anatomy

Root anatomy is similar to that of other monocotyledons and will not be treated here in detail. The outer piliferous layer bearing root hairs is short-lived and a cortical layer of lignified cells forms the boundary layer with the soil. (In Chapter 11 the significance of the root tissues for possible nitrogen fixation is considered.) Within this is an unlignified inner cortex of parenchyma, and an endodermis bearing Casparian strips. This surrounds the stele zone with its xylem and phloem.

Among certain salt-tolerant grasses, the 'excluders' such as *Puccinellia*, the endodermal region is modified by a more pronounced Casparian strip.

Flowering

At its simplest, a stem may lengthen, change over to a reproductive apex and, within as little as three months from germination, complete its life cycle. Maize, growing in the tropics under irrigation can, in this way, yield three generations per year.

Winter wheat and barley establish in late autumn a plant canopy close to the soil surface. Having been subjected to sufficient winter cold to prompt vernalisation during spring and early summer, the stem switches to a reproductive phase preceding elongation. Additionally, from the axils of the basal leaves buds give rise to secondary stems known as tillers each of which establishes its own basal adventitious rooting system.

In contrast with annuals, perennials may persist for several years without appreciable stem elongation but simply adding further tillers, as in the ornamental blue oat (*Helictotrichon sempervirens*), and then eventually flowering.

At the extreme, bamboos may flower after many decades and commonly, though not invariably, manifest holocarpy – not merely of the inflorescence but of the whole plant.

Flowering signifies the reproductive stage and to this we now turn.

Panicles, Spikelets and Florets 4

Two themes are important in this chapter. One relates to the diversity of the panicle and its component parts. The other concerns questions of interpretation.

Eventually, in all but the most persistently vegetative bamboos, a grass plant gives rise to an inflorescence which is often a much-branched structure – the panicle. The panicle branches bear 'spikelets', congested structures within which are found one or more florets. Such florets when pollinated set seed, one per ovary, to be retained and shed with the ovary at maturity as a grain or 'caryopsis'. This simple account, so far as it goes, would apply to grasses generally and in so doing conceals all manner of variation not only in the panicle but, too, in the spikelet and floret.

Bamboos

One might describe a generalised grass panicle, highlight one or two contrasted examples and perhaps mention in passing that bamboos were unusual but (however inadequate present knowledge) bamboos fit in somewhere. If it is not yet convenient, it is none the less stimulating to begin with the diversity of bamboo inflorescences and relate them to what is seen elsewhere. The following considerations are useful.

1. For the hundred or so genera of bamboo the range of inflorescence types is remarkably diverse.
2. Recalling Arber (1934) the adoption of a tree habit with its consequent slowing of generation turnover retards evolution. (Some ephemerals might achieve 150 generations per century while, for certain bamboos, the corresponding figure would be two or three in the same period.)
3. In some grasses, the switch from annual to perennial habit can be under quite simple genetic control.

4. Long generation time in bamboo is not irrevocable and can, for example, be greatly reduced in some cases by *in vitro* culture (Nadgauda *et al.*, 1990).

5. Condensations, sometimes superimposed, appear to have occurred, and recognising the possibility helps interpret the structure of the inflorescence.

Could it be then that, dispersed among different bamboos, are remnants of a 'protopanicle' that we might reassemble? Is it as if, along a stream bed, we found, dispersed by time and storm, a haul of pottery shards from which we might reconstruct the original vessel? Or, are the fragments so eroded that any reconstruction is partial or ambiguous? Changing the analogy, do the perhaps 'slick' inflorescences of 'modern' non-bamboo grasses suggest that bamboos themselves remain in some kind of time-warp?

Other Grass Inflorescences

It should not be thought that bamboos monopolise all that is odd or un-usual among grass panicles. In Panicoideae, for example, if *Coix* and *Zea* are familiar to us we might overlook their extraordinary inflorescences. Even so, the relatively small number of bamboo genera do contain a con-centration of unusual inflorescences. To indicate this, unusual features among bamboos are compared in Table 4.1 to the corresponding situation we find among grasses generally. For example, among bamboos frequently one finds three stigmas per floret. Elsewhere two stigmas is the common situation although in Panicoideae three occur in *Neurachne*. Again, in bam-boos, the genus *Neohouzeaua* lacks both paleas and lodicules, a feature none the less shared with *Alopecurus* and *Cornucopiae* (Pooideae) and *Zoysia* (Chloridoideae) among other non-bamboo grasses. While we assume two glumes per spikelet to be typical, this number varies, conspicuously so, in bamboos but is not unknown elsewhere. With these reservations, we can for comparison recognise a typical 'non-bamboo'.

What is an Inflorescence?

In the lily family, a bluebell having produced some leaves sends up between them a stalk or peduncle on which are shorter stalks or pedicels each bearing a flower. The whole assembly of peduncle, pedicels and flowers comprises an inflorescence of which this particular example would be a 'raceme'.

In grasses the situation is more complex. Could one simply argue that because florets are produced throughout the panicle this latter is an inflorescence? To do so ignores the possible significance of the spikelets.

Table 4.1. Bamboo inflorescences compared with other grasses.

Bamboos	A typical 'non-bamboo'
Panicles	
A great diversity showing typical panicles, globose heads, false panicles and forms with and without spatheoles.	An open-branched structure or with branches more truncated to form a spike or with spikelets joined directly to the axis of the raceme subtended by a spatheole
Spikelet	
One to several florets (sometimes augmented in pseudospikelets)	One to several florets
Glumes 2, 3 *Myriocladus*	Two 'barren' glumes
2–4 *Guadella*	
4–6 *Nastus*	
2–3 + 1–2 lemma-like *Hickelia*	
0 *Schizostachyum*	
Glumes sometimes with potential for pseudospikelet formation	Unsubtending glumes
Glumes sometimes subtended by bracts and/or prophylls	
Floret	
2–3 stigmas (4–6 *Ochlandra*)	2 stigmas
3 or 6 anthers	3 anthers
(120 *Ochlandra*)	
[Beyond bamboos but within Bambusoideae 6 anthers, *Oryza*, 1, 2, 3, 4 or 6 *Ehrharta*]	
3 lodicules (0 *Oreobambos*, 1–15 *Ochlandra*, 0 *Schizostachyum*)	2 lodicules
1 palea (0 *Neohouzeaua*) 1 lemma	1 lemma

Are these really the inflorescences and the panicle merely a branching structure that supports them? This is not just hair-splitting since close examination of the spikelet shows it to be an elaborate structure with, doubtless, a complex evolutionary history of its own.

If then what we had thought of as an inflorescence, namely the panicle, were not one, how can we legitimately import from, say, the Liliaceae terms like 'raceme'? How fussy or precise do we expect to be?

Some ambiguities

Descriptive terms applied by various authors to grass 'inflorescences' (panicles) in their various forms are not particularly consistent. 'Spikes' and 'racemes' can be used to refer to the same structure in one grass and adjectives such as 'spiciform' (in the form of a spike) occur. A spike can be presented as a 'false' spike readily creating the impression that things are not what they seem. Is it possible to clarify the situation and have a system that applies to grasses as a whole?

A clarification

The following assessment, if not accounting for every case, could be expected to have wide validity. Part of its interest is not merely in straightening out the terminology but in raising awareness of evolutionary changes that have probably taken place among the grasses.

1. The panicle is assumed to be the original form of grass inflorescence and is recognisable by its open structure of primary, secondary and even tertiary branches ultimately ending in spikelets. *Oryza*, *Panicum* and *Poa* are examples.

2. The spikelet is seen as a specialised structure, for practical purposes as 'determinate' apart from bamboo 'pseudo-spikelets'.

3. 'Condensation', a curious process of 'compression' and 'simplification', has had two conspicuous effects:

a. it underlies the spikelet in its present form and

b. it has led to inflorescences more resembling a spike than a panicle (spikelets are thus aggregated, sometimes very closely so).

4. Carried to extremes, condensation can lead to a situation like that in *Zoysia minima* where one spikelet is attached to a central axis, all trace of a complex branching panicle system having disappeared. Moreover the spikelet in this case contains a single functional floret.

5. We can recognise a botanical 'sleight of hand', namely to regard the spikelet as a flower (which it is not). This, somewhat dubiously, allows us to regard an inflorescence with a single central axis carrying spikelets joined to it by stalks (pedicels) as a 'raceme'. *Cf.* the example of the bluebell above.

6. If this sleight of hand is ruled out, the term raceme is disregarded and 'spike' is acceptable. *Agropyron* and *Lolium* are examples.

7. If the form of the inflorescence in outline is 'spike'-shaped but close inspection shows it to be made up of short branches densely clothed with spikelets the (not very satisfactory) term 'false spike' is sometimes used. This situation occurs with *Cenchrus*, *Hordeum* and *Pennisetum* among many others.

8. Where the spikelets are crowded on relatively few branches that spread out from a common area of attachment this is a panicle simplified to a 'digitate' arrangement like the fingers of a hand. *Cynodon*, *Digitaria* and *Eleusine* are examples of this.

9. Either parts of one panicle or separate panicles can be committed to one or the other sex. *Buchloë*, *Coix* and *Raddia* exemplify these instances.
10. Fusion of inflorescence branches can lead to 'polystichy', a situation apparently confined to the ear of maize.

To this there are two likely reactions. One would be to propose a rigorous suitably comprehensive new terminology complex enough to account for the many different occurrences. The other, perhaps preferred, is suitably pragmatic recognising that despite its shortcomings the present usage is widely current and likely to remain so.

The Spikelet

Although departures from it occur, both in bamboos and elsewhere, our perception of the typical grass spikelet is of two glumes below one to several florets each of which consists of a lemma, palea, two lodicules, three anthers and two stigmas that surmount the ovary. What must not be overlooked are the remarkable subsidiary detail in structure and the equally impressive diversity of function that occurs in the 'all-purpose' spikelet (Chapman, 1990, 1992b). It is not entirely clear how the (nearly) all-purpose spikelet came to predominate but it does and, in so doing, provides a marvel of adaptation.

Can we, as with the panicle, see any kind of pattern among the different types of spikelet that have evolved? Before attempting this it will be useful to digress and examine two structures. One is peculiar to some bamboos, namely the 'pseudospikelet', and the other, the floret, is general to the Poaceae.

The pseudospikelet

At the base of most grass spikelets is a pair of glumes to which the adjective 'barren' is sometimes prefixed. This is because in their axils can be found neither a floret nor a bud. Above the glumes, each surrounded by the lemma and palea, are one or more florets but the number is normally finite and characteristic of a particular species. On this basis the spikelet is 'determinate' rather than 'indeterminate' or capable of indefinite extension. An interesting exception to this statement occurs with the cultivar 'Kay Muri' of *Eragrostis tef* grown in *in vitro* culture (Tefera, 1992); in place of the eight florets normally formed *in vivo* about 12 were produced in culture. This is part of a wider issue, namely the extent to which artificial culture of grass inflorescences can yield information about developmental pathways. For a review see Chapman (1995).

Some bamboos differ from this in that the glumes subtend a bud capable of becoming a flowering branch thus augmenting the original spikelet. Three striking features are:

1. The first structure of the new branch is a 'prophyll' (considered in detail later).
2. A glume is differentiated with a bud in its axil.
3. This latter bud can again elaborate a further flowering branch with a prophyll as *its* first organ.

In contrast, therefore, with the spikelet, the pseudospikelet is indeterminate. The obvious question is which came first, the spikelet or the pseudospikelet? We could assume either that the latter was the basic type abandoned in favour of the simpler spikelet by most grasses or that the pseudospikelet is a later innovation adopted by some bamboos in response to, say, infrequent flowering and where every growth point in the reproductive structure is pressed into service for flower and fruit formation. The repeating pattern in the pseudospikelet is said to be 'iterauctant'. Pseudospikelets occur for example in *Bambusa, Dendrocalamus, Dinochloa, Gigantochloa, Hickelia, Melocalamus, Melocanna, Ochlandra, Oreobambos, Oxytenanthera, Schizostachyum* and *Thyrsostachys*, all bamboos of tropical or subtropical distribution.

The floret

Among the grasses, the small flower is normally referred to as a floret and taken to include lodicules, stamens and an ovary surmounted by stigmas. From a strict botanical viewpoint matters can be improved by adding to these two structures – the palea immediately below the floret and the lemma below the palea. Palea, lemma and floret comprise the 'anthoecium'. Above the (normally two) glumes of a spikelet there is thus one or more anthoecia. Such rigour is seldom evident, a spikelet being said to contain florets of which palea and lemma are assumed to be included. This is the approach taken here therefore and links with the literature that is most relevant.

If the assumption is made that grasses had lily-like ancestors, florets of some bamboos show most evidence of this. It is then possible to devise a radiating series to show connections with other grasses. Incorporating such 'trends' into various spikelet arrangements indicates how present day diversity might have resulted.

Figure 4.1 illustrates the principal events that might be conjectured, proceeding as follows.

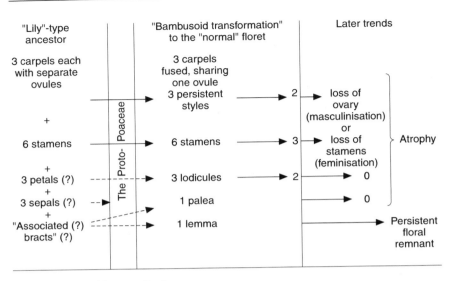

Fig. 4.1. Evolution of the grass floret.

1.   Invoking a purely hypothetical group, the 'Proto-Poaceae', provides a link between some lily-like ancestor and the grass family.

2.   The legacy of the Proto-Poaceae to the Poaceae could be the compound ovary merging three carpels and a single shared ovule, six stamens, three lodicules and, from whatever source, providing the palea and lemma. (They, especially the palea, seem underivable from sepals.)

3.   There then followed the 'bambusoid transformation' whereby progressive loss of one of the three styles, half the stamens and one of the lodicules established the 'normal' or 'modern' or 'economy' grass floret. Scattered among present-day bamboos particularly are surviving instances of the earlier condition together with those showing 'partial modernisation'. Thus:

Sasa	3 lodicules	6 stamens	3 styles
Guadella	3 lodicules	6 stamens	2 styles
Melocanna	2 lodicules	6 stamens	2 (or more) styles
Merostachys	3 lodicules	3 stamens	2 styles

a trend completed in modern grass with

| | 2 lodicules | 3 stamens | 2 stigmas |

4.   This modernised form is widely represented among the remaining subfamilies but is capable of further modification. Three common manifestations are

   a.   Masculinisation where the ovary becomes non-functional and may be absent or present in only partially developed form.

b. Feminisation where the anthers are non-functional or absent.
c. Loss of both types of sex organ and perhaps loss of palea and lodicules – everything in fact except the lemma. This comprises 'atrophy', the lemma serving to indicate the presence of a once-complete, or 'perfect', floret.

This scheme should be recognised as a guide to which there are curious exceptions. *Dendrocalamus* and several related genera lack any lodicules. In one unusual bamboo, *Brachyelytrum*, there are only two stamens. And perhaps most extraordinary of all, *Ochlandra* is a genus of bamboo where lodicules can vary from 1 to 15, stamens 15 to 120 and stigmas 4 to 6.

Clifford (1961) took the bamboo *Arundinaria* floret as a theoretical starting point and showed how other grass florets could be derived from it – primarily by loss of parts – and makes the interesting point that given known 'trends' the *Arundinaria*-type model could generate more derivatives than are known to exist among the grasses. Clifford also drew attention to the curious fact that if six stamens were ancestral within a tropical forest situation (and perhaps with entomophily) it is difficult to explain three rather than six in a wind-pollinated context. He suggests that retention of the panicle and the development of automatic self-pollination as in cleistogamous types (discussed below) might offer insurance mechanisms. Given these various trends in the floret is it now possible to set out some convincing scheme of spikelet evolution?

The evolution of the spikelet

The spikelet is best regarded perhaps as 'enigmatic'. What exists today is the outcome of response to selection in which 'condensation', that amalgam of compression and loss of parts, is a consistent theme. The likely stages could well have been:

1. Variation for bamboos in glume number between 0 as in some species of *Oryza* to as many as 6 in *Nastus*, but settling down eventually to 2 in most grasses.
2. (Controversially) the establishment of progressive loss of function either upward or downward through the spikelet. The assumption is that a spikelet contained 'several' florets. In pooid grasses loss of function has been from the tip of the spikelet toward the glumes. Conversely in panicoid grasses there has been loss of function upward from the florets nearest the glumes. See for example the discussion of *Sorghum* later in this chapter.
3. Where loss of function has proceeded only to the point where one or other sex function has become defunct, other associated modifications of the spikelet can occur. In *Buchloë*, which is dioecious for example, the

glumes in males and female plants are strikingly different being larger and more ornamented in the female. In *Zea*, which is monoecious, the spikelets in the tassel containing only male florets are relatively unmodified except for the near absence of female parts. Those of the female by contrast are so modified, in modern maize, as to barely conform, at first sight, to the notion of spikelet.

4. Integral to the evolution of the spikelet are all those modifications that relate to distribution and survival. In *Stipa barbata*, for example, the lemma has evolved a plumed extension that assists wind dispersal of fruits (Fig. 4.2). In *Aristida funiculata* (Fig. 5.1) the spiky awns of the lemma cause thousands of fruits to be aggregated into a ball perhaps 30 cm in diameter that will roll before the wind across the desert and fragment caryopses from it on impact as a means of dispersal.

5. In genera such as *Pennisetum* a cluster of spikelets is surrounded by hairs interpreted as the remnants of former branches of the panicle – again, seen as a consequence of condensation. The same is the case with *Saccharum* illustrated in Chapter 11.

The prophyllum or prophyll

A leaf and a lemma have a recognisable midrib or 'keel' causing the latter, for example, to be boat-shaped. There does, however, occur in various places on a grass plant a two-keeled structure, the prophyllum or prophyll. Its occurrence is not random but both orderly and thought-provoking.

It occurs (a) at the origin of a vegetative branch, (b) within the pseudospikelet at the origin of a flowering branch, (c) in its most common manifestation as the palea subtending a floret, (d) rarely, among panicle branches as in the very unusual bamboo *Glaziophyton* subtending individual spikelets and (e) in its most readily seen form as the outer (first-formed) husk of a maize ear (Fig. 4.3).

The prophyll of any new branch stands with its back toward the earlier one from which it originated. Arber (1934) drew on her acquaintance with heraldry and from 'adossé' derived the word 'addorsed' (backing onto) to describe the orientation of the prophyll. That the prophyll is two-keeled has been ascribed either to the fusion of two separate structures or because of indentation by the originating structure but the matter is unresolved.

The prophyll is followed by one or more single-keeled structures, either leaves, glumes as in *Glaziophyton* or lodicules in most grasses. If the prophyll is equivalent to the palea why not assume a similar equivalence among leaves, glumes, lemmas and lodicules? While that might be too extreme it throws into prominence that deliberately equivocal statement made earlier about 'associated bracts' in Fig. 4.1. Because the palea is

Fig. 4.2. *Stipa barbata*, an example of a plumed seed (this example is about 40 cm long).

Fig. 4.3. (a) From left to right: vertical stem, two-pointed prophyllum and maize ear wrapped in remaining husks. (b) The terminal portions of the prophyllum and the first true husk compared in maize.

two-keeled and because the prophyll occurs at various places on a grass plant, it is unsatisfactory to equate it with the sepal of a lily-type ancestor. By extension, need we automatically regard the lodicules *above* the palea as petal equivalents? Could they instead be modified bracts that have misled us to think of them as petals simply because in so many bamboos they occur in threes?

The lemma

The leaf of flowering plants, the angiosperms, is capable of great modification being in some cases reduced to scales devoid of photosynthetic function. If it subtends a flower in its axil it is referred to as a bract but often its photosynthetic function persists and physiologically it can be regarded quite simply as a leaf.

In grasses the leaf is divisible to sheath, ligule and lamina. For some species the proportions of successive leaves on a culm vary considerably. Leaving aside the prophyll therefore, the glumes, leaves and lemmas are recognisably but variants on a theme. The glume is equivalent to a leaf sheath and can sometimes produce a foliaceous tip equivalent to the lamina, as in *Phyllostachys aurea* for example. A similar situation is often evident in the husks of a maize ear.

By equating the body of the lemma with the leaf sheath the awn, where it exists, can be seen equivalent to the lamina. This process in reverse can be seen convincingly in 'hooded' barley mutants. Here the awn is transformed into a winged green lamina curved into a 'helmet' shape.

Having established the lemma's credentials, the sometimes spectacular modifications of which it is possible then fall into place. *Stipa barbata* has already been mentioned and Fig. 4.4 illustrates *Stipagrostis ciliata*.

The awn is considered in more detail in Chapter 5.

'Trends'

If a set of different inflorescences were arranged in this or that order they could be presented as giving evidence of a trend (which can of course be read backwards or forwards or rearranged with perhaps additional material). From such a process has any convincing trend begun to emerge?

When culms, inflorescences and florets are examined and single prophylls occur could we argue that each is evidence of a recurrent theme? This is that each prophyll-bearing structure can progressively lose parts and the culm, panicle, spikelet and floret all yield evidence of a series of such branches in various states of reduction (or occasionally expansion).

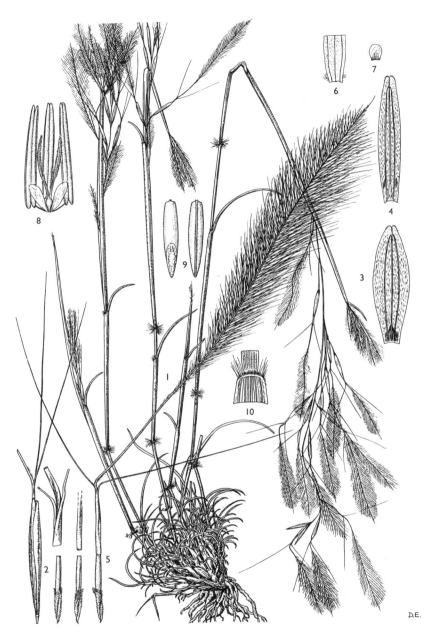

Fig. 4.4. *Stipagrostis ciliata* (Desf.) de Winter.
1. habit × $\frac{2}{3}$; **2.** spikelet minus central branch of the awn × 3; **3.** lower glume × 3; **4.** upper glume × 3;
5. floret × 3; **6.** lemma × 3; **7.** palea; **8.** flower × 6; **9.** grain × 6; **10.** ligule × 2.

On this basis, the basal leaf and the prophyll would form lemma and palea with the lodicules derived from leaves above the palea. Similarly, 'glumes' bearing an axillary bud that could generate a flowering branch with prophyll would be a component of the pseudospikelet. Where this happens the glume is not barren but fertile and perhaps should be called a lemma. If next we see such a lemma and prophyll subtended a spikelet in determinate fashion then the spikelet is present on a branch recognisably reduced. Where the lemma has a non-functional bud it is 'transitional' and, when this has gone, the structure is a glume.

Wide variation exists in the lemma in regard to texture, varying from papery to coriaceous, in differences between the members within a spikelet and in the degree of ornamentation present.

The glumes

Below the lemmas are a pair of glumes, sometimes more among the bamboos. Their function is largely protective. Like the lemma they can be awned or unawned but in the former case do not approach extremes of plumed development seen in some lemma awns. They can vary in size, being contrasted members of a pair as in *Hickelia* for example, or, as in barley, not separable into upper and lower glumes as they stand side by side. They may be shed with the fruit, as in *Eragrostis*, or be retained, as for example in *Cynodon*. Additionally, as in the upper glume of a female *Buchloë* spikelet or the lower glume of *Trilobachne*, they may be symmetrically trilobed. Curiously, a hyaline patch on the upper glume of *Thyridolepis* occurs on the lemma of the closely related *Neurachne*.

Further details of floret and spikelet parts are given in the various botanical drawings of different species that occur throughout this book.

The lodicules

Three lodicules immediately below the anthers in bamboo normally prompts botanists to assume their equivalence to petals. Even however in those rare examples of grass entomophylly such as *Pariana* the lodicules do not provide for insect attraction. Elsewhere among some bamboos and most other grasses only two lodicules are found per floret and these are small, inconspicuous structures whose sole function seems to be hydroinflation, so as to separate the palea and lemma and permit floret opening.

What can be explained?

As indicated earlier, the Bambusoideae provides various contrasts that can be arranged in a series to show, for example, the 'modernisation' of the grass floret. The conclusion is that above just a pair of glumes is a lemma and palea enclosing two lodicules, three anthers and a pistil, that is an ovary with one ovule and surmounted by two stigmas. The unit comprises a demonstrably efficient structure and every ingredient appears to fulfil some role. Even when, as in *Buchloë*, for example, the lodicules are larger in the male florets than the female ones, this is readily explained. Those in the male inflate to part the palea and lemma to permit efficient anthesis. Such a process is unnecessary in the female since the stigmas protrude through the slit between the palea and the lemma. Consider however the cases listed in Table 4.2 where omission of glumes, palea or lodicules occur.

Among these instances, of the substantial number that might have been included, seemingly key ingredients are shown to be 'expendable'. Even so, the lemma is almost invariably present. Andropogonoid spikelets carry this to an extreme where the lower floret in a spikelet is non-functional but, none the less, the now seemingly useless lemma persists. For detail see the concluding part of this chapter dealing with *Sorghum*. Again, there are apparently no cases of simultaneous omission of both glumes and palea.

In *Oryza* the barely discernible remnants of the glumes lie below a robust lemma and palea that enclose the single floret. The remarkable success of rice reminds us that a seemingly 'incomplete' floret is able to function with, apparently, high efficiency.

While at a functional level, the various omissions are not readily explained, they none the less provide convenient characters for the taxonomist. The absence or near absence of glumes helps distinguish the tribe Oryzeae.

Table 4.2. Selective omission (−) of spikelet parts.

Genus	Glumes	Palea	Lodicules
Alopecurus	+	− (or v. small)	+
Coleanthus	−	+	−
Cornucopiae	+	−	−
Heteropogon	+	−	−
Jouvea	−	+	−
Lygeum	+	+	−
Neohouzeaua	+	−	−
Oryza	−	+	+
Zoysia	one (or two very unequal)	present or absent	−

Anemophily and departures from it

Throughout the angiosperms entomophily or insect pollination is common-place and it is the departures from it that prompt comment. Even so, greater familiarity shows that insect visitors can be very diverse and some-times so inconspicuous as almost to defy detection. Again even in a single genus such as the dicotyledon *Ruellia*, species occur variously adapted to moth, bee and bird pollination and, while flower shape, colour and size vary appropriately, the groundplan is virtually unaltered. If, as has been conjec-tured, grasses had some kind of 'lily-like' ancestor, it could have been entomophilous. Notwithstanding this, virtually no trace of entomophily in the grass flower form remains and, in terms of *function*, pollinating insects are rare visitors.

There are exceptions and Bogdan (1962) presented a detailed account of tropical grasses being visited by bees in the genera *Apis* and *Nomia*. That pollination *can* be effected by these insects seems likely but this does not amount to dependence upon such a means. In maize, bees can be seen collecting pollen but, given the spatial separation of male and female inflorescences, the consequences for pollination are probably negligible.

Fig. 4.5. *Brachiaria decumbens*, an instance of prolifery (or vivipary).

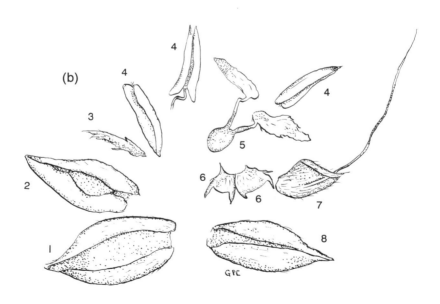

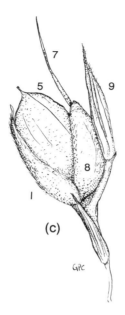

Fig. 4.6. (*opposite*) (a) Sessile spikelet of *Sorghum* dissected. (b) Key to sessile spikelet.
1. lower glume; **2.** lower lemma; **3.** upper palea; **4.** anthers; **5.** ovary → caryopsis; **6.** lodicules; **7.** upper (awned) lemma; **8.** upper glume. (c) Spikelet pair in *Sorghum* at maturity. Numbers as in Fig. 4.6(b), plus **9.** pedicellate spikelet.
Notes to Figure 4.6(a), (b)
i. Items 1–8 comprise the sessile grain-bearing *spikelet*.
ii. 1 and 8, the lower and upper glumes, subtend two florets.
iii. The lower floret is represented only by 2, the lower lemma.
iv. Items 3, 4, 5, 6 and 7 comprise the upper (grain-bearing) *floret*.
Notes to Figure 4.6(c)
v. It can be seen that the sessile grain-bearing spikelet is accompanied by the pedicellate spikelet.
vi. Items 2, 3 and 6 have been compressed by the enlarging grain.
vii. The anthers have been discarded, as have the stigmas surmounting the ovary which is now a recognisable caryopsis.

Are there any instances where insect pollination of grasses is essential? A detailed study by Söderstrom and Calderon (1971) concentrated upon the more herbaceous bamboos in the genera *Olyra* and *Pariana* mentioned earlier growing in the understorey of tropical forests in South America.

Such an environment is shaded and has minimal air movement. These investigators presented a list of insects associated with grass florets of these species pointing out that still air was inappropriate to anemophily and that the insect genera *Chauliodontomyia* (Cecidomyiidae – gall midges) and *Pericyclocera* (Phoridae – hump-back flies), for example, were consistently associated with *Pariana* even in widely separated localities, together with numerous other insects whose occurrence might be more incidental. Söderstrom and Calderon drew attention to the bright yellow anthers of *Pariana* and their concentration in spike-like inflorescences and suggested they functioned to attract insects.

Apart from such examples should we conclude anemophily to be the norm? There remains another alternative, namely that through 'cleistogamy' the grass flower is closed until after self-pollination has occurred. Cleistogamy is found in barley, rice, sorghum and wheat as the basis for automatic self-pollination and self-fertilisation. Since the mechanism is nearly but not completely 100% efficient, a low or very low level of outcrossing can occur and be the basis for genetic change over time in a population.

Again it should be recognised that while pollination can stimulate seed reproduction it does not automatically follow that normal fertilisation will occur. Apomixis is a pollen-induced parthenogenesis whereby a plant gives rise to exact copies of the mother plant; it is considered subsequently in Chapter 9. Finally, the spikelet might undergo a modified development so as to take the form of small plantlets. Such prolifery (or vivipary) bypasses the sexual system completely, see Fig. 4.5.

Spikelet modification: an instructive instance

Sorghum bicolor belongs to the andropogonoid grasses where characteristically there is a mixed spikelet arrangement. Sessile grain-bearing spikelets are accompanied by pedicellate (stalked) spikelets normally reduced to a pair of glumes and little else. An obvious question to ask is whether the pedicellate spikelets, currently serving no apparent use, might not be made grain-bearing, perhaps thus putting evolution into reverse. Such instances have been described but they are both inconsistent in inheritance, and in practice make for no sustainable, or even detectable, increase in yield. Figures 4.6a, b and c show dissection of the sessile spikelet and the arrangement of spikelet clusters. Restoration of function to the pedicellate spikelet is an issue relevant to maize evolution and is discussed in Chapter 12.

It is now appropriate to consider how grass reproductive structures function.

Generation and Dispersal

5

A grass panicle normally has a distinctive appearance recalling not only the family but quite often a particular genus. Maize is the obvious example but, in ways not readily described, the panicles of *Cynodon*, *Dactylis*, *Festuca*, *Lolium*, *Oryza* and *Poa*, for example, tend to be almost instantly recognisable. Given its relatively short period of development, the homogeneity of its parts and its holocarpy (whether or not this latter extends to the rest of the plant), it is convenient to regard the entire panicle as a functional unit. *Pennisetum alopecuroides* (Chinese fountain grass) varies the pattern somewhat having both restricted seed formation and an extended flowering season and, as a result, is adopted as an ornamental. More typically, grass panicles, when ripe, fragment to create 'diaspores', the structure dispersed to establish the next generation. As will be shown later, a diaspore, depending on species, is a variable structure consisting of the caryopsis alone or with particular associations of chaff. Again, it underlines the remarkable versatility of the spikelet.

Pollination

A point of contrast is between some bamboos that have been studied closely and other grasses. Usually in grasses, the lodicules inflate and force apart the palea and lemma, the anthers protrude and shed their pollen, and this easily recognised stage is called 'anthesis'. Grasses tend to be protandrous, pollen shedding preceding stigma receptivity, but protogynous exceptions occur such as *Pennisetum* where conspicuous style exsertion is ahead of anthesis. Anemophily, wind-borne pollen, is the rule, as was shown earlier for grasses, the pollen grains being captured among the trichomes of the stigma.

For two bamboos, *Ochlandra travancorica* (Venkatash, 1984) and *Dendrocalamus strictus* (Nadgauda, pers. comm.), copious stigmatic secre-

71

tion is recorded and it is unknown how far this property extends among bamboos. It is a matter of considerable interest since the remainder of grasses are reckoned to have 'dry' stigmas (Heslop-Harrison and Shivana, 1977). In this remarkable mechanism dry pollen rapidly attaches to the stigma of a plant with which it is compatible taking time measured in seconds – for example *Cenchrus ciliaris* (28–45), *Chrysopogon aucheri* (19), *Panicum maximum* (10), *Pennisetum squamulatum* (17) (all Ibrahim, pers. comm.), *Secale cereale* (90) (Heslop-Harrison, 1979).

Watanabe (1955) described how, after pollen attachment, moisture is briefly exuded from the grain and resorbed as a prologue to the emergence of the germ tube and its entry into the stigma. For a review see Heslop-Harrison (1987) where the grass pollen stigma interaction is placed in the context of other plant systems.

For self-pollination to be automatic or nearly so requires self-compatibility. Among the grasses such a condition is possibly the result of a collapsed, pre-existent two-gene self-incompatibility system shown to occur in some outbreeding grasses and apparently unique to the Poaceae.

Self-incompatibility

Among many flowering plants prevention of self-fertilisation is under genetic control. Although there are variations in detail the systems are either 'gametophytic' where the pollen is, as to its incompatibility reaction with the stigma, autonomous, or, alternatively, pollen reaction is to some extent conditioned by the plant from which it arose in which case the system is 'sporophytic'. In grasses, the system is gametophytic but differs in being under the control of not one major gene as is generally the case, but two to give the so-called S–Z system.

The essentials of the grass system are that there are two loci, S and Z, which are unlinked and therefore segregate independently. Both S and Z consist of multi-allelic series $S_{1,2,3...}$, $Z_{1,2,3...}$. A diploid grass would therefore possess two S and two Z alleles, for example $S_1S_2Z_1Z_2$. Style tissue is diploid and therefore possess all four alleles while the pollen grain possesses one allele from each gene.

When pollen lands on the stigma, the incompatibility reaction is conditioned by whether or not the pollen grain shares alleles with that stigma and style. If it shares none or one, the pollen will be compatible. If it shares both, it will be incompatible. Table 5.1 indicates which pollen grains will function when assorted genotypes of pollen parent, sharing different numbers of alleles in common, are crossed to the female parent $S_1S_2Z_1Z_2$. Pollen genotypes which share both alleles in common with those present in the style, and which would therefore fail, and shown by †. For the three crosses of Table 5.1 which do produce offspring, the possible offspring genotypes are shown in Tables 5.2 to 5.4.

The following points occur:

1. Pollen having only one allele (or none) in common with the stigma will function.

2. A zygote can be homozygous at the S locus or at the Z locus but it cannot be homozygous simultaneously at both loci, e.g. $S_1S_1Z_2Z_3$ or $S_2S_3Z_1Z_1$ but not $S_2S_2Z_3Z_3$. Most zygotes are heterozygous at both loci.

3. The greater the allelic differences between two plants, the greater the likelihood that pollen will function. Within a population this mechanism greatly increases the probability of successful cross-pollination, since the number of cross-compatible genotypes is related to the product of the number of alleles for each gene. Most estimates have identified some tens of alleles for each gene, resulting in hundreds of cross-compatible combinations.

4. Grass pollen is often produced in vast quantities and is in that sense 'expendable'. The ovary produces one ovule only and is thus a valued resource. It is worth commenting that on the basis of incompatibility reactions no ovule is wasted though, of course, for other reasons, such as lack of physiological support or some kind of damage, it may fail.

S–Z incompatibility is unusual in that the incompatibility reaction is retained in polyploids. Many problems about the grass incompatibility system remain unanswered. They include the following. We do not know whether the S–Z system is unique to grasses or whether all self-compatible

Table 5.1. A model two-gene incompatibility system showing the fate of pollen from the designated parent genotype landing on a style of genotype $S_1S_2Z_1Z_2$. The † symbol indicates pollen grain genotypes which fail.

	(A)	(B)	(C)	(D)
Pollen parent:	$S_1S_2Z_1Z_2$	$S_1S_2Z_1Z_3$	$S_1S_3Z_1Z_3$	$S_3S_4Z_3Z_4$
Pollen grains:	S_1Z_1†	S_1Z_1†	S_1Z_1†	S_3Z_3
	S_1Z_2†	S_1Z_3	S_1Z_3	S_3Z_4
	S_2Z_1†	S_2Z_1†	S_3Z_1	S_4Z_3
	S_2Z_2†;	S_2Z_3	S_3Z_3	S_4Z_4
Result:	All alleles common: all pollen fails	3 alleles common: half of pollen fails	2 alleles common: a quarter of pollen fails	No alleles common: no pollen fails

Table 5.2. Genotypes resulting from the pollinations shown in column B of Table 5.1.

	Functional pollen genotype	
Embryo sac genotype	S_1Z_3	S_2Z_3
S_1Z_1	$S_1S_1Z_1Z_3$	$S_1S_2Z_1Z_3$
S_1Z_2	$S_1S_1Z_2Z_3$	$S_1S_2Z_2Z_3$
S_2Z_1	$S_1S_2Z_1Z_3$	$S_2S_2Z_1Z_3$
S_2Z_2	$S_1S_2Z_2Z_3$	$S_2S_2Z_2Z_3$

Table 5.3. Genotypes resulting from the pollinations shown in column C of Table 5.1

	Functional pollen genotype		
Embryo sac genotype	S_1Z_3	S_3Z_1	S_3Z_3
S_1Z_1	$S_1S_1Z_1Z_3$	$S_1S_3Z_1Z_1$	$S_1S_3Z_1Z_3$
S_1Z_2	$S_1S_1Z_2Z_3$	$S_1S_3Z_1Z_2$	$S_1S_3Z_2Z_3$
S_2Z_1	$S_1S_2Z_1Z_3$	$S_2S_3Z_1Z_1$	$S_2S_3Z_1Z_3$
S_2Z_2	$S_1S_2Z_2Z_3$	$S_2S_3Z_1Z_2$	$S_2S_3Z_2Z_3$

Table 5.4. Genotypes resulting from the pollinations shown in column D of Table 5.1.

	Functional pollen genotype			
Embryo sac genotype	S_1Z_3	S_3Z_1	S_3Z_3	
	S_3Z_3	S_3Z_4	S_4Z_3	S_4Z_4
S_1Z_1	$S_1S_3Z_1Z_3$	$S_1S_3Z_1Z_4$	$S_1S_4Z_1Z_3$	$S_1S_4Z_1Z_4$
S_1Z_2	$S_1S_3Z_2Z_3$	$S_1S_3Z_2Z_4$	$S_1S_4Z_2Z_3$	$S_1S_4Z_2Z_4$
S_2Z_1	$S_2S_3Z_1Z_3$	$S_2S_3Z_1Z_4$	$S_2S_4Z_1Z_3$	$S_2S_4Z_1Z_4$
S_2Z_2	$S_2S_3Z_2Z_3$	$S_2S_3Z_2Z_4$	$S_2S_4Z_2Z_3$	$S_2S_4Z_2Z_4$

grasses have become so by degeneration of S–Z. It is not clear why polyploidy leaves the system intact (as compared with gametophytic systems elsewhere) nor has any significant progress been made in a molecular understanding of the pollen–stigma interaction for grasses. Where a self-incompatible grass has been studied in detail the S–Z system has been found to operate, but while therefore self-incompatibility is known in many genera its detailed S–Z understanding is confined to eight species, all of them pooids. Given the wet stigmas found in at least two bamboos and the ability now to truncate the life cycle by *in vitro* culture, there is not only interest but opportunity to resolve the nature of self-incompatability in that subfamily. For a detailed review of S–Z incompatibility in grasses see Hayman (1992).

Pollen and Gene Travel

Although pollen is normally airborne, it need not travel far. For *Pennisetum glaucum* and *Zea mays* on occasion less than 10% of shed pollen is found more than 5 m from source. By contrast, for *Lolium perenne* and *Phleum pratense* more than 20% of pollen travelled more than 200 m from source (Richards, 1990). Such observations are of course conditioned by aspect, wind speed and rainfall. What is more to the point is how far genes borne by pollen actually travel since this is the biologically significant consequence of anemophily. It is possible to show that grasses tolerant to heavy metals can convey such tolerance by pollen many metres beyond the contaminated soil for which such a property is necessary but it does not follow that metal tolerance will spread beyond where it is useful. McNeilly and Antonovics (1968) compared two grasses, *Agrostis tenuis* (copper tolerant) and *Anthoxanthum odoratum* (lead tolerant), growing on mine spoil with their nearby non-tolerant counterparts growing in pasture free of these metals. For each species it was possible to obtain tolerant × non-tolerant hybrids experimentally. None the less, those growing on mine spoil flowered about a week earlier precluding much gene flow on to non-tolerant plants. There is a curious twist to this work. Using non-tolerant *Anthoxanthum* as the female parent, crossing pollen on to it from tolerant plants was unsuccessful. If one makes the assumption that selection pressure *for* tolerance is greater on mine spoil than selection against tolerance in pasture the effect is greater resistance to tolerance coming off the mine spoil, perhaps the opposite of expectation.

If a grass is outbreeding and self-incompatible, but a particular clone has colonised a wide area and excluded competitors, the pollen dispersed will be, in practice, wasted. For these and other reasons alternative grass life-styles are considered in detail in Chapter 9.

Intergeneric Crosses

Can outcrossing be 'too wide' – involving partners in different species? For the well-studied British flora, Stace (1975) records that about 20 grass genera form interspecific hybrids. Of these genera the following make authentic intergeneric hybrids: *Festuca* × *Lolium* (× *Festulolium*), *Festuca* × *Vulpia* (× *Festulpia*), *Ammophila* × *Calamagrostis* (× *Ammocalamagrostis*, latterly *Calammophila*) and *Agrostis* × *Polypogon* (× *Agropogon*). Of these generic hybrids all are sterile or nearly so and none involves more than one tribe.

Artificial hybrids

It is appropriate to distinguish between hybrids surviving naturally and those for which all manner of physiological support is available under laboratory conditions. Watson (1990) provided a world list of grass intergeneric hybrids. Chapman (1990) examined the margins of crossability, in particular with a view to finding the 'widest' hybrids that had been reported and how such claims might be authenticated. One can define a set of widening criteria – within or between subtribes, within or between tribes, within or between subfamilies – and then arrange various claims by different authors on this scale. Three such examined by Chapman (*ibid.*) involved crosses between subfamilies. These were *Oryza* (Bambusoideae) × *Leptochloa* (Chloridoideae) (Farooq and Naqvi, 1987), *Oryza* × *Pennisetum* (Panicoideae) (Wu and Tsai, 1963) and *Oryza* × *Sorghum* (Panicoideae) (Zhou *et al.*, 1981). These claims are, to say the least, surprising, but what matters is evidence. In these and similar claims there need to be voucher herbarium specimens, photographs, mitotic and meiotic analyses, and, nowadays, some evidence of distinctive DNA from each parent being represented in the offspring.

Perhaps the 'widest' hybrid for which, at the time, convincing evidence was produced was a *Pennisetum* (Paniceae) × *Zea* (Andropogoneae) hybrid (Nitsch *et al.*, 1986). The hybrid, significantly, required embryo rescue, showed chromosome loss and serves to underline the precarious viability of intertribal crosses.

Gametogenesis

Sexual reproduction in grasses is typical of many angiosperms. There is a Polygonum-type eight-nucleate embryo sac arising from the single surviving megaspore produced at meiosis. For a detailed treatment of the grass ovule and its constituent structures see Greenham and Chapman (1990).

The most puzzling feature of the grass embryo sac is the prolificity of the antipodal apparatus. While for most flowering plants three antipodal cells occur, in grasses these may either multiply or show massive nuclear replication through polyteny. In *Sasa paniculata*, for example, 300 antipodals were reported (Yamura, 1933). In *Hordeum vulgare* Cass and Jensen (1970) found up to 100 antipodals, and in *Chloris gayana* the three antipodal cells remain but enlarge through chromatin replication so as to occupy half or more of the gametophyte volume (Chikkannaiah and Mahalingappa, 1975). Antipodals are thought of as 'ergasto-plasmic', that is, metabolically very active in view of their ultrastructure which shows generous endoplasmic reticulum, numerous mitochondria and dictyosomes. What is far from clear is what, for all their seeming activity, they actually do. They are present at a key stage in the life of the plant, namely around the time of endosperm and embryo initiation and are transitory. Greenham and Chapman (1990) proposed that their function could be either 'secretory' or 'sacrificial' (or of course both). Are they therefore somehow crucial in the establishment of endosperm? If this is so, how is it that they are absent from the apomictic embryo sacs of grasses (to be discussed later) where, clearly, they are in no sense crucial?

Fertilisation

Following pollination, compatible pollen will germinate and a pollen tube penetrate the stigma and style moving between cells toward the micropyle. Pollen under the control of S–Z genes and found to be incompatible will not germinate; pollen tubes may enter the stigma but falter thereafter.

So important both to the plant itself and to plant breeders and geneticists is this phase of the life cycle that it is helpful to identify key stages and comment on features of interest – both the known and those about which we yet seek knowledge.

1. Style penetration. In maize, the food reserves of the pollen grain are inadequate for its growth through 20–25 cm of style. Nutrition is available from the style but the mechanisms involved are not well understood.

2. Modification of pollen tube contents. Grass pollen is trinucleate. The two male cells deriving from the generative cell are each retained *within* the vegetative cell cytoplasm. This situation persists in the germ tube after pollen germination but in grasses the two male gametes are not intimately linked with each other.

3. Cytoplasmic diminution. On their journey through the pollen tube each male cell produces cytoplasmic extensions into which mitochondria and plastids enter, the extensions then being 'pinched off' (Mogensen and

Rusche, 1985). Loss of mitochondria and plastids (each containing DNA) means that the likelihood of cytoplasmic inheritance via the male gamete is apparently diminished in contrast to the situation in the female cytoplasm.

4. Micropyle entry. The pollen tube eventually enters the micropyle. Recently, the cells at or near the micropyle in some grasses have been shown to have a specialised ultrastructure and form the so-called 'embellum' (Busri et al., 1993). Conceivably the functions of such a structure are to recognise, nourish or guide the pollen tube or to conduct some or all of these.

5. Synergid entry. The pollen tube enters the degenerating synergid via the filiform apparatus and discharges its contents. For grasses, this does not include the vegetative nucleus which is retained within the pollen tube. The two male gametes at this stage each appear to consist of a nucleus enclosed in a small amount of cytoplasm that still contains some mitochondria plastids and other organelles, the whole being enclosed by a plasma membrane but lacking a wall. For a detailed discussion see Mogensen (1990).

6. Sperm dimorphism. The elegant researches of Russell (1986) established for Plumbago zeylanica that the two male cells differed in their mitochondria:plastid ratio and that a preponderance of plastids was most often associated with the sperm entering the egg. These findings stimulated a search among other plants for a similar dimorphism. Knox and Singh (1990) took the matter to its logical conclusion arguing that if one cell of male origin were consistently to fuse with the central cell it was not properly called a sperm but an 'associate cell'. The evidence for morphological dimorphism for the grass cells is currently equivocal but less easily detected biochemical differences may exist in the membrane boundaries that are involved in selective recognition mechanisms (Mogensen, ibid.; Knox and Singh, ibid.).

7. Embryo sac relations. Our understanding for grasses at present is that the male protoplast interacting with the egg contributes only a nucleus, the cytological debris being left on the egg surface. The male protoplast interacting with the central cell completes protoplasm fusion, the nucleus and cytoplasm of both participants being involved (Mogensen, ibid.). The zygote and subsequently the resulting embryo develop in close association with the endosperm, a tissue to which the male parent has also contributed. Although in general terms the endosperm is said to 'nourish' the embryo, the matter is, seemingly, extremely subtle as reference to the next subsection will show.

8. For many years fertilisation in vitro with isolated male and female gametes has been commonplace in zoological research. The plant parallel, double fertilisation in vitro leading to functioning zygotes eludes us and such partial results as we do have are equivocal. Kranz et al. (1991a,b) isolated enzymically embryo sacs and eventually eggs and sperms of maize.

Fusion could only be achieved with the aid of an electric shock. Moreover the resulting 'quasi zygote' could divide but would not differentiate. Since the male sperms had not passed through either the pollen tube or the synergid, had some equivalent of 'capacitation' been omitted? Because an endosperm was not present was some essential contribution from this direction lacking? We do not know and the area remains crucial for investigation.

9. Differentiation. The zygote, under normal circumstances, begins to differentiate as it proceeds through embryogenesis. For grasses the zygote is one cell that is self-evidently totipotent. Transform this and maybe the whole resulting plant will be transformed. Unlike the Solanaceae for example, among grasses very few cells (other than the zygote) are readily totipotent for reasons at present largely obscure.

10. Omitting double fertilisation. Hitherto, it might seem as if double fertilisation were an essential prerequisite to differentiation. Many grasses are however 'apomictic'. Only one fertilisation occurs, namely that involving the central cell and in any case the effective embryo sac is one arising not from the archesporium but from the nucellus. Recalling the work of Kranz *et al.* (*ibid.*) there *is* an endosperm present (and, it will be noticed, of bi-parental origin) but the egg it 'nourishes' although unfertilised can function as if it were a zygote and undergo embryogenesis.

That a nucellar cell can expropriate some, but not all the functions of the archesporially derived sexual embryo sac comprises a tantalising problem in grass biology with, as will be shown later (see Chapter 11), implications for plant breeding.

Zygote to Embryo

Nuclear fusions occur first involving the egg and then in the central cell. Except among polyploids, a diploid zygote is in close association with a triploid endosperm. Despite their shared genomes only the former differentiates but diploidy does not offer the *sine qua non* of differentiation since haploid, triploid and higher ploid embryos can all complete their differentiation. In the Poaceae the ability of a cell to differentiate a whole organism (totipotency) had seemed unique to the zygote in the presence of endosperm. Although exceptions are known, this remains the norm and provides a severe problem for tissue culture using grasses. The situation is changing as techniques improve and recently scutellar tissue has been made embryonic (Nehra *et al.*, 1994) for example.

Although among various grasses embryogenesis will differ in detail, the pattern described for rice (in Table 5.5) is broadly characteristic of the family.

Table 5.5. A timetable for rice embryo/endosperm development. The ovule contains a single flask-shaped embryo sac about 150 μm in length and with 10–15 antipodals on the side toward the placenta. The egg cell is about 25 μm in length enclosed by synergids. A 'pocket' of the central cell containing the egg apparatus has projecting wall ingrowths and becomes an area of active metabolism. Pollination to fertilisation takes about one hour. One synergid degenerates before the other, the former providing the route for pollen tube entry. After fertilisation, the nucellar tissue near the antipodals shows disintegration (based on Jones and Rost, 1989).

Time after anthesis	Embryo	Endosperm
	Note: Ontogeny proceeds from the chalazal end but storage proceeds in the opposite direction	
4 hours	An unequal transverse division of the zygote establishes a larger suspensor initial toward the micropyle and a smaller embryo initial toward the chalaza	
1 day	Suspensor has divided longitudinally to give two cells. Embryo consists of 5–8 cells	Free nuclear division stage in progress
2 days	Embryo globular, both synergids disappear	Free nuclear at chalazal end, cellular at micropyle end
3 days	Embryo consists of 100–200 cells. Polarity evident. Starch accumulation at the base	
4 days	Scutellum distinct, coleoptile becoming evident. Radical not organised. Starch throughout to the coleoptile	Degenerating layer around the embryo
5 days	Scutellum elongating, coleoptile over-arching the shoot apex, epiblast evident. Radicle developing. Lipid bodies detectable	
6 days	All organs recognisable that will be present in the mature seed. Protein storage first detectable	
7 days	Protein accumulated in the coleoptile	
8 days	Connection to nucellus via the suspensor broken. Aleurone continuous except where embryo connected to nucellus	

Embryogenesis: Considerations of Fine Structure

The events preceding, during and after double fertilisation continue to absorb the attention of many botanists since they raise questions about the changeover from gametophyte to sporophyte, ontogeny and the enigmatic significance of the endosperm. The advent of electron microscopy and various developments in microanalysis have stimulated further interest and its relevance to grain fill in cereals is at once obvious.

Given the embryo, endosperm and nucellus, which feeds what and for how long and by what means? An ovary in which is an ovule with its embryo sac is far smaller than the caryopsis into which it will develop. While therefore the nucellus could nourish the embryo and the endosperm this could only be short-term and then the nucellus is overwhelmed and disappears. The endosperm, if it is adequately to serve the embryo, has itself to be generously supplied. However, given the demise of the nucellus, could the developing endosperm itself be a nutrient conduit to the embryo at the same time?

Schell *et al.* (1984) found evidence in maize of endosperm modified to 'transfer' cells in the placentochalazal region. Typically, such cells have a dense cytoplasm with endoplasmic reticulum and wall projections. Highly ergastoplasmic cells were found, too, at the base of the embryo (around the suspensor). The implication is that these two parts of the endosperm represent respectively input and output areas of the nutrient supply to the embryo. Jones and Rost (1989) for rice show disintegration of the nucellar cells adjacent to the antipodals implicating these in obtaining nutrients from those cells. As in maize, the endosperm tissue around the base of the embryo also disintegrates. For wheat, Smart and O'Brien (1983) reported non-vacuolate endosperm cells around the base of the embryo but do not attribute any role to the antipodals.

More recently, Engell (1994) in an ultrastructural study of barley antipodals has shown the following features. The number of antipodal cells present is between 35 and 50. They remain uninucleate but the nucleus enlarges eventually about 50 hours after pollination to about five times the original size (presumably by polyteny). The cells form a fan-shaped array pointing toward the placenta. After pollination, the antipodals eventually vacuolate. Plasmodesmata can occur between antipodal cells but their cell walls abutting the nucellus never have plasmodesmata and the walls are mostly thin but thickened with wall invaginations. Remnants of nucellar cells are pressed together opposite the antipodals. Antipodal cells show ergastoplasmic cytoplasm and disappear as the endosperm cellularises.

There is a contrasting case, perhaps helpful, to discuss for *Pennisetum*. *P. glaucum* ($2n = 28$) in artificially induced tetraploids will hybridise with *P. squamulatum* ($2n = 54$), an obligate apomict, provided this latter is the pollen parent. The F_1 hybrid with 41 chromosomes has been the focus of much attention as a possible means of creating apomictic pearl millet, see for example Dujardin and Hanna (1988). The embryo sacs of *P. squamulatum* lack antipodals and the problem then becomes one of exploring that part of the embryo sac where, by comparison with *P. glaucum*, the antipodals would have been. Put simply, does the central cell, lacking antipodals, become modified to re-create their function? Chapman and Busri (1994) reported that for asexual embryo sacs there are no central cell wall projections where the antipodal cells *would have been*. By contrast, in *sexual* embryo sacs when antipodals are present these latter cells, the antipodals, *do* have wall projections where they abut the nucellus. The obvious though unproven conclusion seems unavoidable, that if antipodals are present they serve a transfer function, perhaps, and if not the plant makes adequate alternative arrangements.

The 'antipodal enigma'

Despite the evidence of numerous studies of 'ergastoplasmic' antipodal cells and their hallmarks of transfer cells their function is unknown. Even our most obvious assumption must take account of the fact that apomictic embryo sacs manage without them. In addition there are two old observations of antipodal hypertrophy in hybrids by Brink and Cooper (1944) and Beaudry (1951) to which cytologists have repeatedly drawn attention. The basic assumption in both cases is that hybridity impedes normal antipodal performance which in turn disrupts endosperm establishment and thus sterility ensues. The assumption, though convenient, is unproven, and, given advances in electron microscopy, appropriate for careful reappraisal.

Caryopsis Diversity

Eventually, the caryopsis comes to maturity consisting of embryo and endosperm within a seed, itself tightly contained within the hardened caryopsis. The caryopsis is a focus of major agricultural interest deserving a book to itself. In lieu of this the reader is referred to Rost and Lersten (1973), which provide a synopsis and bibliography in which 11 characters (entire caryopsis, development, germination, homologies, embryo, endosperm, aleurone layer, caryopsis coat, starch, aleurone transfer cells and protein bodies) were scored by the authors for 25 genera.

After embryogenesis

The impression might be that following completed embryogenesis, with an endosperm food reserve, germination would follow shortly. This overlooks not only the onset of dormancy but the remarkable extent to which the tissues surrounding the caryopsis have developed so as to facilitate its protection and distribution. The common assumption that the chaff hardens and the lemma extended into an awn provides a means of distribution overlooks the remarkable number of variants there are on this theme. Clayton (1990) remarked:

> The reduction of the spikelet to one floret followed by reinstatement of a many-flowered unit by aggregating spikelets seems paradoxical, but its effect is to bring novel structures into play. Sometimes these augment existing trends but sometimes they seem to be restoring functions discarded by earlier adaptations.

The implications of this are discernible both in the inflorescence and in subsequent fragmentation at its demise. Table 5.6 presents in simplified form a range of alternatives. For a more detailed treatment see Clayton (*ibid.*).

One of the most curious means of seed dispersal is the 'grass-ball' found, for example, in *Aristida funiculata* mentioned briefly in Chapter 4. The mechanism is as follows. A large sward of *Aristida* growing under semi-desert conditions sheds a unit consisting of palea, caryopsis and lemma – this last having a three-branched awn. The glumes are retained by the plant. The shed plant parts are swirled by the wind and form an interlocking mass that can aggregate beyond the size of a football, becoming a quite dense mass. In stronger wind the grass ball is rolled across an open landscape. In

Table 5.6. Grass diaspores.

Dispersal unit	Genus	Significant features
Whole inflorescence	*Leptaspis (Scrotochloa)*	The inflorescence breaks off as a whole unit
	Spinifex	The plants are either male or female. Female inflorescence shed entire, male inflorescence sheds spikelet separately
	Zea	Whole (female) inflorescence not dispersible – due to human selection
Groups of spikelets	*Tristachya*	A triad of three spikelets having fused pedicels are shed together
Whole spikelet	*Sorghum*	The functional (sessile) spikelet is shed together with the remnant of the non-functional (pedicellate) spikelet
Whole anthoecium	*Hordeum*	The unit in 'normal' barley includes palea and lemma adherent to the caryopsis
Caryopsis only	*Triticum,* 'naked' *Hordeum*	Cultivated wheat which is free threshing. Barley mutants with non-adherent palea and lemma
Seed only	*Sporobolus* (some)	The caryopsis modifies to extrude a mucilage-coated seed
Multiple dehiscence	*Catalepis*	Rachis fragments after which the spikelets are then separately deciduous
'Adherent' dispersal	*Aristida* (some)	A 'grass ball' is formed (see text)

transit, the repeated jolts detach from the ball units consisting of caryopsis, palea and *lower* portion of the lemma. The upper portion, the awn, remains embedded in, and comprises the bulk of, the grass ball. One of these balls, shown in Fig. 5.1, was given to Kew under the erroneous impression that it was a fallen weaver bird's nest (T. Cope, pers. comm.).

The awn

While still on the plant in the inflorescence prominent awns can offer some protection against bird damage. Given a choice, for example, between awned and unawned *Sorghum*, there is some evidence that *Quelea* birds prefer the latter (Doggett, 1988). Since the awn is photosynthetic (Grundbacher, 1963) there was a vogue for breeding it into commercial wheat for a time, but demonstrable gains in yield were minimal and the practice is now largely discontinued, at least for higher latitudes. The notion that the awn is hygroscopic and will drill the caryopsis into the ground is widely current and will now be examined.

The almost automatic assumption that a twisted hygroscopic awn will, on being moistened, drill its caryopsis into a soil crevice has been challenged in a series of papers (Peart, 1979, 1981, 1984; Peart and Clifford, 1987). Among the results of this work are the following. Spikelets

Fig. 5.1. Grass balls of *Aristida funiculata.* The scale is a 30 cm ruler.

can fragment or not to provide structures, the diaspores, that may have hygroscopic (active) awns, passive (rigid) awns or both or, alternatively, the diaspore may lack awns. These authors point out that not merely the awn but also other structures condition the effectiveness of the diaspore, notably the 'callus', a pointed structure at the base with ranks of backwardly pointing hairs. The role of the hygroscopic awn is probably effective primarily in propelling the seed *laterally* across a surface toward (by chance) a suitable microsite such as a crevice. Evidence for 'drilling' was lacking except for *Heteropogon contortus* and then only under unusual circumstances. The passive awns, if present, together with the callus functioning as a set of spring-loaded barbs will manoeuvre the diaspore into a partly or completely buried position and, if the structure is also upright, its chances of germination and establishment will be enhanced. Such a mechanism operates, for example, in *Aristida vagans* and *Microlaena stipoides*. Awned diaspores tend to germinate promptly while unawned types more often undergo dormancy. The authors explain this by pointing out that awned types occur on cracking soils while unawned ones are more common on hard-setting sandy surfaces. Passive burial is a time-consuming process and prompt germination would be counterproductive. It is also suggested that dormant buried seeds can be stimulated to germinate by fire.

Peart (1979) draws attention to an evolutionary convergence whereby callus on *Schizachyrium fragile* develops from the rachis internode while that on *Danthonia tenuior* originates on the glume of the sterile pedicellate spikelet. Unless the callus or some other structure provides firm anchorage for a diaspore at or near the surface, pressure by the emerging radicle could dislodge the structure and establishment would fail. See Fig. 5.2.

Dispersal

Whatever the nature of the diaspore, there remains the problem of dispersal. What is the point of a long-distance dispersal mechanism that takes a diaspore beyond the ecological range of the species? Clayton (1990) suggests that the bulk of the fruits move less than 10 m from the parent though with occasional individuals travelling much greater distances. Adaptations to animal transport include secretion of oil in an elaiosome (*Yakirra*), sticky secretion (*Oplismenus*) and spines (*Aristida*). Many grasses when ingested have a portion viable after being excreted. Ridley (1930) made a detailed compendium of plant dispersal mechanisms and there are numerous references there to grasses. Antelopes, ants, birds, bison, cattle, ducks, elephants, horse, mammoth (extinct), marmot, pigs, red deer, reindeer, rhinoceros, sheep, termites, water buffalo and yak are all considered as means of distributing particular grasses. Mammoth food, for example,

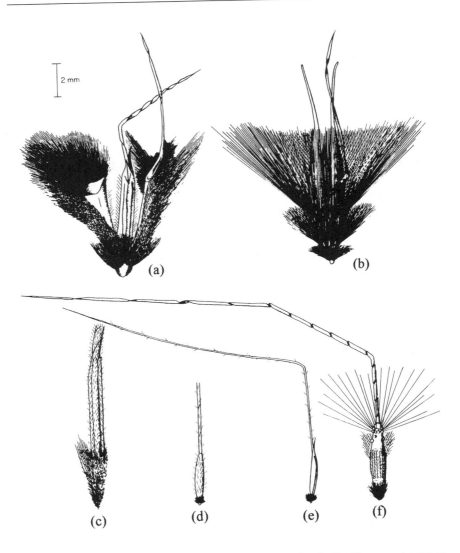

Fig. 5.2. Contrasted grass diaspores (from Peart, 1979) (a) *Schizachyrium fragile* with one passive awn on the pedicellate spikelet, a hygroscopic awn on the fertile spikelet, and dense retrorse hair on the basal callus and elsewhere. (b) *Danthonia tenuior* with two passive awns developed from the lobes of the lemma, a single hygroscopic awn and dense retrorse hairs on the basal callus and elsewhere. (c) *Heteropogon contortus* with no passive awns, a stout hygroscopic awn and retrorse hairs on the basal callus and elsewhere. (d) *Stipa verticillata* with no passive awn, one hygroscopic awn and retrorse hairs largely confined to the basal callus. (e) *Dichelachne macrantha* with no passive awns, one hygroscopic awn and retrorse hairs confined to the basal callus. (f) *Dichanthium sericeum* with no passive awns, one hygroscopic awn, various sizes of retrorse hairs but especially dense on the basal callus.

although not confined to grasses, included *Agropyrum cristatum* and *Hordeum violaceum*.

Grasses are opportunist though not uniquely so and many owe distribution beyond their original confines to humankind. The 'Africanisation' of the New World grasslands was referred to in Chapter 1 and is reconsidered in Chapter 10.

Ridley's account is useful partly because he was an acute botanical observer moving about before the modern era of mass transport and partly because his account is so readable. He describes for example the Oxford ragwort, *Senecio squalidus*, taking train journeys and, of a grass growing on Christmas Island, he remarks with a touch of irony – 'on Mount Ophir, at the spring where drinking water is obtained, and the transport coolies wash their clothes, I once found the common widely dispersed American *Paspalum conjugatum*.'

Of 'piano grass', *Themeda arguens*, so-called in Jamaica, it is alleged that a civil servant in the days of the British Empire was relocated from south east Asia to the West Indies and lined the packing case of his piano with this grass which subsequently naturalised there. He thus appears to have enriched not only the musical life of his new posting but its botany as well.

Dormancy

A simple assumption is that for a ripe caryopsis the likelihood of its germinating would diminish with time and that its viability would be better maintained under dry cold conditions than moist warm ones. Such an assumption, though not unreasonable, has in practice to be subject to qualification. The caryopsis itself, for example, is not merely viable but possessed of a particular style of metabolism that can vary among species and be subject to some degree of environmental conditioning. The familiar situation is a series of events from shedding, through dormancy, its relief through after-ripening and, then, subsequently germination. If, having completed after-ripening, the seed is deprived of some factor for germination it may after a period of quiescence enter a second dormancy. There is an obvious link to be made between this and a seasonally imposed control of germination indicating the evolution of economy of seed use. A recent example is the contrasted behaviour of *Bromus tectorum* (cheatgrass) an annual, commonly germinating in the autumn, and *Leymus cinereus* (basil wild rye) a perennial germinating in spring. These grasses inhabit similar regions of western North America but have different establishment and seed production strategies (Meyer *et al.*, 1995).

The examples in Table 5.7 of various mechanisms are primarily to dissuade the reader from generalisation. Some of these examples, with

Table 5.7. Dormancy and germination among grasses.

Aegilops geniculata	Caryopses nearer the base of the spikelet have higher germination percentages. Mother plants raised in short days give caryopses with lower germination percentages	Gutterman (1992)
Avena fatua	Seeds matured on the plant at 20°C have lower dormancy than at 15°C	Peters (1982)
	Secondary dormancy can be induced by anoxia in imbibed seeds	Symons *et al.* (1987)
Dactylis glomerata	Light requirement for germination declines during storage	Probert (1992)
	Dormancy is more evident in N European rather than S European populations	Probert (1992)
Echinochloa crus-galli	Light dependency for germination can be abolished by 1/2 hour at 46°C in 37% of individuals	Taylorson and Di Nola (1989)
Echinochloa turnerana	In a population individuals differ widely in their photon requirement to promote germination	Ellis *et al.* (1986)
Festuca pratensis var. *apennina*	Populations at lower altitudes generate seeds that require lower optimum temperatures than do those from higher altitudes	Linnington *et al.* (1979)
Hordeum spontaneum	Removal of endosperm near the embryo and/or caryopsis layers covering the embryo assists germination	Gutterman (1992)
Oryza glaberrima	The longer and warmer the storage environment the greater the loss of dormancy	Ellis *et al.* (1983)
Polypogon monspeliensis	The longer the daylength to which the mother plant is exposed, the more rapid is seed germination	Gutterman (1992)
Zizania palustris	Chilling of submersed seed breaks dormancy as can removal of lemma and palea or layers covering embryo	Probert (1992)

others, are discussed in detail by Murdoch and Ellis (1992) and Probert (1992). What is apparent, on a species by species basis, is that an ecotype is harmonised to its particular environment in a way that hardly permits extrapolation to other cases. And if there is to be successful occupancy of some new habitat, among the first descendants of the founder population selection will presumably function upon dormancy mechanisms as on other aspects of the species life-style. For a detailed account of dormancy in grasses see Simpson (1990). Following the breakage of dormancy and germination the events set in train are those described in Chapter 3.

Taxonomy 6

Taxonomy is the oldest aspect of plant science for which there is a technical literature and traces its origins from Theophrastus (370–287 BC), a pupil of Aristotle. Theophrastus produced ΠΕΡΙ ΦΥΤΩΝ ΙΣΤΟΙΑΣ – *Enquiry into Plants* based on the flora of Greece augmented by plants introduced from the overseas campaigns of Alexander the Great. Table 6.1 indicates the grasses, converted to modern genera, of which Theophrastus was aware and provides a means by which to assess the work of later taxonomists. In some ways the list is surprising, including two bamboos (*Bambusa* and *Dendrocalamus*) and three species of wheat (*T. aestivum*, *dicoccum* and *monococcum*). Theophrastus was also aware of regional and varietal differences among wheat and barley for example. Curiously, although he knew of *Lolium temulentum* as a nuisance among cereals, *Lolium perenne*, common enough in Greece, appears to have escaped his notice.

The next writer of consequence for which we have a record is Dioscorides writing in the first century AD. His list of grasses is shorter but his interest was primarily medical.

Thereafter, little creative thinking about plants of which we have a record occurred in Western Europe until the Renaissance. There were traditions in Arab, Chinese and Indian writing but their impact on mainstream taxonomy has been indirect.

The Renaissance is generally said to have begun in 1401 in Florence with the competition for design of the Baptistery doors of San Giovanni and for which Ghiberti and Brunelleschi were the conspicuous participants. It is only fair to point out, however, that the displacement of Arab scholars westward by disruption of Islamic centres of learning was one significant contributory cause. Thereafter, first in Italy but soon after throughout much of Europe, human curiosity was seemingly reborn and botany together with almost every other kind of enquiry began to benefit. Columbus' voyages with their introduction of plants from the New World, perhaps echoing the effect of Alexander's travels, exposed the inadequacies of tradition and stimulated new interest. Many botanists contributed to the new learning but one of the most interesting was John Ray, an English parish priest, in some ways Linnaeus' most immediate predecessor. Ray helped define the idea of a species and in so doing laid the foundations for subsequent controversy about how far they were permanent productions of the Creator (for a discussion see Raven, 1950). The addition of *Coix*, *Sorghum* and *Zea* in the *Methodus Plantarum* (1682) is a glimpse of the world opening up to botanical exploration. Ray in a later work, *The Wisdom of God Manifest in the Works of Creation* (1704), drew attention to the remarkable productivity of wheat when grown in a range of widely different climates.

Linnaeus 'the prince of botanists' of course brought order and system that need not delay us here since they are so widely celebrated elsewhere. Although Linnaeus is often credited with belief in the fixity of species, he demonstrates, from a closer acquaintance with his work, that this was not so. Stearn (1957) has shown how Linnaeus' views altered and how toward his later years he was, in all but name, an evolutionist and this a century before Darwin. How far Theophrastus' concept of a genus might concur with that of Linnaeus is of course speculative, quite apart from our being sure that his Greek names can, in every case, be converted to the appropriate binomials. Even so it is evident that some at least of Theophrastus' ideas survive 2000 years later in the work of Linnaeus and beyond. And although Linnaeus included *Pharus* and *Olyra*, grasses of bambusoid affinity, in his writings he did not do so for *Bambusa* and *Dendrocalamus* known to his Greek predecessor.

The Nineteenth Century

Botanical exploration, partly stimulated by Linnaeus and partly by imperialist ambitions, enhanced European herbaria, and at Kew, George Bentham and Sir Joseph Hooker brought that establishment to pre-eminence. In *Genera Plantarum* (1883) Bentham and Hooker set out a tribal arrangement for grasses and Hackel (1890) contributed, via the *Pflanzenfamilien* of Engler and Prantl, a not dissimilar one (see Table 6.1).

During the nineteenth and early twentieth centuries there were numerous treatments of the Gramineae of which Table 6.1 provides only a sample. Others include Harz (1880–82), Lamb (1912), Bessey (1917), Schellenberg (1922) and Hayek (1925).

Table 6.1. An outline of the historical development of grass taxonomy.

Author	Taxa
Theophrastus *Enquiry Into Plants*	*Aegilops, Andropogon, Arundo, Avena, Bambusa, Calamagrostis, Cymbopogon, Cynodon, Dendrocalamus, Hordeum, Imperata, Lolium, Oryza, Panicum, Phragmites, Polypogon, Saccharum, Setaria* (see Fig. 6.1), *Triticum* (3)
Dioscorides First century AD *De Materia Medica*	*Aegilops, Arundo, Avena, Cenchrus, Cynosurus, Erianthus, Hordeum, Lolium, Oryza, Panicum* (3), *Phalaris, Phragmites, Triticum*
Ray (1682) *Methodus Plantarum*	'Gramina'. *Aegilops, Alopecurus, Avena, Festuca, Lachryma-jobi* (i.e. *Coix*), *Hordeum, Lolium, Melica, Milium, Oryza, Panicum, Phalaris, Secale, Sorghum, Triticum, Zea*
Linnaeus (Note: Genera from *Genera Plantarum* (1767) Species number () from *Species Plantarum* (1753))	Gramineae. *Aegilops* (5), *Agrostis* (12), *Aira* (14), *Alopecurus* (4), *Andropogon* (12), *Anthoxanthum* (3), *Apluda* (1), *Aristida, Arundo* (9), *Avena* (12), *Bromus* (11), *Brixa* (4), (*Briza*), *Cenchrus* (5), *Cinna, Coix* (2), *Cornucopiae, Cynosurus* (9), *Dactylis* (2), *Elymus* (5), *Festuca* (11), *Holcus* (7), *Hordeum* (6), *Ischaemum* (2), *Lagurus, Lolium* (2), *Lygeum, Melica* (3), *Milium* (2), *Nardus* (4), *Olyra, Oryza, Panicum* (20), *Paspalum, Phalaris* (5), *Pharus, Phleum* (4), *Poa* (17), *Saccharum* (2), *Secale* (4), *Stipa* (3), *Tripsacum, Triticum* (7), *Uniola* (2), *Zea, Zizania* (2)

Thereafter, numerous other genera were described and by the time of Hackel (1890) had risen beyond 300. Taxonomists by then had begun to group genera into 'tribes' as indicated below.

Author	Taxa
Bentham and Hooker (1883)	Tribes arranged in two series: Series Panicaceae Series Poateae
Hackel (in Engler and Prantl, 1890)	**Gramineae** Tribes 1. Maydeae, 2. Andropogoneae, 3. Paniceae, 4. Oryzeae, 5. Phalarideae, 6. Agrostideae, 7. Aveneae, 8. Chlorideae, 9. Festuceae, 10. Hordeae, 11. Bambuseae.
Hackel (in the 1896 English translation)	i. Maydeae, ii. Andropogoneae, iii. Zoysieae, iv. Tristogineae, v. Paniceae, vi. Oryzeae, vii. Phalarideae, viii. Agrostideae, ix. Aveneae, x. Chlorideae, xi. Festuceae, xii. Hordeae, xiii. Bambuseae.
Bews (1929)	**Gramineae** **Pooideae** 1. Bambuseae, 2. Pharea, 3. Festuceae, 4. Aveneae, 5. Chlorideae, 6. Hordeae, 7. Agrostideae, 8. Zoysieae 9. Phalarideae, 10. Arundinelleae, 11. Oryzeae. Panicoideae 12. Melinideae, 13. Paniceae, 14. Andropogoneae, 15. Maydeae.
Avdulov (1931)	**Poatae** *Series* Phragmiformes Tribes — Arudineae, Oryseae, Stipeae, Centotheceae Bambuseae *Series* Festuciformes Tribes — Festucaceae, Frumentaceae *Series* Sacchariferae Tribes — Sporoboleae, Chlorideae, Andropogoneae, Tristagineae, Maydeae, Paniceae
Prat (1960)	Subfamily Festucoidees (8 tribes)* Subfamily Panicoidees (4 tribes)* Subfamily Chloridoidees (3 tribes)* Subfamily Bambusoidees (1 tribe)* Subfamily Orysoidees — Olyroidees (Pharoidees) (2 tribes)* (Phragmitiformes) (4 tribes)* Genres a position discutée (some 15 genera)

Current classifications recognise about 40 tribes representing more than 750 genera and 10,000 species. The adoption of 'series' by Bentham and Hooker foreshadowed today's subfamily groupings and in addition to Panicoideae and Pooideae we recognise Arundinoideae, Bambusiodeae and Chloridoideae together with a disputed smaller subfamily Centothecoideae as indicated earlier in Chapter 2.

One other change has been the partial adoption of the family name 'Poaceae'. 'Gramineae' is sanctioned by the International Code of Botanical Nomenclature even though it is not based on a legitimate included genus. The later name Poaceae (based on *Poa*) is an allowable alternative.

* Plus clusters of uncertain affinity.

The Twentieth Century

Bews (1929) expanded upon the work of his nineteenth century pre-
decessors but using essentially a similar approach. A most significant new
development was that due to Avdulov (1931) who introduced a cytogenetic
interpretation of grass taxonomy. He provided comparative mitotic metaphase illustrations for many genera and hundreds of chromosome counts in
an essentially evolutionary setting.

That perceptive student of grasses, Agnes Arber (1934) wrote:

> ...for – now that we are becoming aware of the part which hybridisation has
> apparently played in evolution – we have to disembarrass ourselves of 'the
> notion that a form is *either* a species or a hybrid'. The new conceptions of race
> constitution, which have arisen out of genetics and cytology, are not easy to fit
> into the orthodox systematic framework; but the old disharmony between the
> standpoint of the field and herbarium worker, and of the genetical cytologist, is
> rapidly being resolved by co-operation in research.

Arber then in a footnote refers to C.D. Darlington who subsequently
through three decades bestrode international genetics as a brilliant
polemicist. His insight combined with a most caustic wit and a certain
cultivated irreverence ensured the penetration of genetics into every area of
biology – including grass taxonomy (see Darlington, 1932, 1937, 1956,
1969; Darlington and Janaki-Ammal, 1945; Darlington and Mather, 1949;
Darlington and Wylie, 1955).

What Arber foresaw emerged as 'experimental taxonomy', of which a
conspicuous product was G.L. Stebbins' (1950) *Variation and Evolution in
Plants* considered subsequently in Chapter 11.

Eventually two new ingredients were added to biology in general,
namely, refinements in biochemistry that led to a better understanding of
proteins and, after a time, to rapid DNA sequencing, together with the rise
of computer science that facilitated the handling of large data sets.

Within the study of grasses an important contribution was that of
Söderstrom. Although he died aged only 51 he left some 70 papers among

---

Fig. 6.1. (*opposite*) A modern representation of *Setaria glauca* (L.) P. Beauv, one of about 20 grass genera
known to Theophrastus.
1. habit × 1; **2.** spikelet × 6 (note bristles interpreted as remnants of branches); **3.** lower glume × 12;
4. upper glume × 11.5; **5.** lower lemma × 11.5; **6.** its palea × 11.5; **7.** upper floret, ventral aspect × 11.5;
8. upper lemma × 11.5; **9.** its palea × 11.5; **10.** flower × 10; **11.** grain × 8; **12.** ligule × 3.
In present-day terms, *Setaria* belongs to the subfamily Panicoideae, tribe Paniceae. The spikelet contains
two florets, the lower one much reduced and non-functional, the upper one hermaphrodite and grain bearing.
Setaria glauca is a cosmopolitan plant of waste ground and is readily eaten by stock.

which was an enthusiastic but scholarly concern for bamboo science. For an assessment of his work see Söderstrom *et al.* (1987).

Contemporary Grass Taxonomy

Two publications set the scene for this discussion. These are Clayton and Renvoize (1986) and Watson and Dallwitz (1988). The former published from Kew is a stimulating reworking of grass genera arranged as 40 tribes in six subfamilies. Its approach is phylogenetic and various groups of grasses are presented in 'bubble' diagrams to indicate their likely relationships. The latter is in some ways quite different, being first available as a database from which, on the basis of arrays of characters or their alternatives, it was possible to assemble clusters of genera. These were eventually presented as diagrams that matched them to five subfamilies and a series of supertribes and tribes. Each system was presented in summary form to permit ease of comparison as Watson (1990b) and Clayton and Renvoize (1992). Finally, the Watson and Dallwitz database was published as a book in 1992. In this version the authors base their treatment on 496 characters per genus.

Modern systems compared

On first acquaintance, comparing the respective summaries, it is the similarities of arrangement rather than the differences that are apparent although these latter are instructive (see Chapter 2). Manipulation of the compendious database is not simply mechanical. For example, Watson (1990) remarks on 'the (unacceptable) clustering of *Hordeum* with *Avena* rather than *Triticum*', clearly symptomatic of recognising the need to exercise continual taxonomic 'judgement'. Judgement is integral to any classification, of course, and Renvoize and Clayton (1992) refer to the methods used as 'as much art as a science' while recognising the options offered through the computer. The traditional approach is sometimes described as 'neural' (Kellog and Campbell, 1987). Renvoize and Clayton (*ibid.*) then discuss characters in terms of taxonomic usefulness indicating that:

1. Photosynthetic pathways have become important.
2. Embryo structure, though difficult to use, is valuable at subfamily level.
3. Fusoid cells are virtually confined to Bambusoideae.
4. Microhairs when modified are good predictors for the tribe Eragros-
    tideae and the subfamily Pooideae.
5. Lodicules can be useful at subfamily level.
6. Large chromosomes where n = 7 characterise the Pooideae.
7. Woody culms characterise the tribe Bambuseae.

8. A barren floret below a fertile one can be diagnostic for the subfamily Panicoideae.

9. Trimerous flowers have phylogenetic implications.

10. Ligules are mostly too erratic to be of much significance.

In general, more emphasis is placed upon reproductive than vegetative characters by taxonomists since the former are less subject to environmental influence. Excluding 10 above it is noteworthy here that six valued features are vegetative.

Given the exercise of skilled judgement it can be possible to recognise a group of grasses as highly distinctive, an instructive example being the tribe Stipeae with seven genera agreed as belonging there and some others placed there by one but not the other system. Even so, Clayton and Renvoize consider the Stipeae as a tribe in the subfamily Pooideae while Watson and Dallwitz ascribe it to the Arundinoideae.

An important point of which to be aware is that sometimes a similar tribal name, for example Arundineae, is differently regarded. For Watson (1990) it includes three genera only (*Arundo*, *Phragmites* and *Thysanolaena*) while for Clayton and Renvoize (1986) it embraces no fewer than 40. A separate tribe, Thysanolaeneae, these authors then recognised was subsumed under (their) Arundineae in their partial revision (1992).

One feature of some interest is the recognition, only by Clayton and Renvoize, of the small subfamily Centothecoideae originally defined by Söderstrom (1981). Placed by Watson and Dallwitz in the Bambusoideae, Renvoize and Clayton (1992) speak of its leaf blade anatomy as 'quite distinctive and bears no relation to the Bambusoideae'. Certainly, if one places some reliance upon embryo structure the presence of a mesocotyl helps set it further apart from the Bambusoideae. It remains an interesting difference between the two systems.

At the level of practical utility, Clayton and Renvoize provide a set of traditional 'keys' down to generic level based essentially on the principle of the 'excluded middle' that originated with Theophrastus. Watson and Dallwitz dispense with this, offering a computerised sorting from characters supplied in any order to identify to genus. A caution is appropriate. It is possible to sort for (say) genera with species lacking glumes and *Oryza* (rice), for example, would be included. If, however, the sterile lemmas there were interpreted as glumes, as is sometimes the case, then rice would 'wrongly' have been included. The current interpretation is more convincing that small projections below the sterile lemmas are rudimentary glumes. Again one might search for florets lacking a palea where *Saccharum* and *Sorghum* would then emerge. As will have been evident from the concluding section of Chapter 4, it is important to recognise that their spikelet arrangement is complex and within which are two florets, one with and one without a palea. The user therefore needs to attend to the detail of the various characters available. Neither Clayton and Renvoize (1986) nor Watson (1992) is a 'field volume'. Their roles here are primarily for professional taxonomists revising local or regional Floras. It is these latter books, complete with waterproof covers, that accompany the field botanist on his or her excursions and permit identification to species level.

Phenetics, Cladistics and Phylogeny

Phenetics refers to observable character states that, via the computer, might show clustering – a process that can be conducted devoid of any evolutionary assumptions. Cladistics, by contrast, claims to reveal likely evolutionary relationships. How do such seemingly contrasted approaches affect grass systematics? Watson (1990) remarked:

> For what it is worth this practitioner subscribes to the view that valid cladistic inferences from taxonomic data of a traditional kind depend upon prior recognition of phenetic groups.

In a later contribution Hattersley and Watson (1992), while expressing appropriate reservations, take a phylogenetic view of the diversification of C_4 photosynthesis.

Kellogg and Campbell (1987), using data from presented alternative cladistic interpretations of the five major subfamilies, find the Arundinoideae the most problematic. Perceptively, the authors comment that phylogeny reconstruction will be seen as a good modelling exercise, a way of examining the consequences of whatever assumptions are made about character distributions and taxonomic relationships.

The foregoing discussion has centred on the 'taxonomy' of grasses but at a technical level it is necessary to recognise three important strands – classification, identification and nomenclature. Much of what has been said so far concerns classification. If plants are allotted to groups, families, genera and species this has little practical value unless specimens can be recognised reliably. The process of identification, utilising dichotomous keys or computer sorting of characters referred to earlier, facilitates this. Finally, the name given to a particular plant or group of plants is of little use unless such names are somehow generally agreed. By reference to words either Latin, or latinised, and used according to the International Code of Botanical Nomenclature, plant names are adopted and validated according to an agreed procedure. In essentials, taxonomy is a science with its origins in Western philosophy although nowadays scientific contributors to it do so from many countries.

An interesting sidelight on conventional taxonomy is available from a study of bamboos in the Yunnan province of south-west China made by Wang (1990) and Wang *et al.* (1993).

Alternative approaches to bamboos in Yunnan

Among the richer countries of Western Europe and North America, bamboos are seen primarily as ornamentals. In mostly poorer tropical countries, bamboos occur spasmodically among other vegetation and can provide garden canes and also serve other uses. *Bambusa vulgaris* growing on steep tropical hillsides can, for example, impede erosion. *Guadua* in Latin America is a relatively large bamboo that can be utilised structurally. An interesting situation occurs where a rich diversity of bamboos coincides with an innovative human tradition to exploit them. While this might describe, say, the southern half of China generally, it reaches its most extreme expression almost certainly in Yunnan, a province in south-western China toward Burma. Here, in the region Xishuangbana, is a remarkable diversity of bamboo species that have long been exploited by various ethnic groups that inhabit the region, including Dai, Ha'ri, Bulang, Yao, Wa, Jinno, Lisu, Miao and others. The bamboos present can be summarised as follows; (Wang *et al.*, 1993): *Bambusa* (9 spp.), *Cephalostachyum* (2), *Chimonobambusa* (3), *Chimonocalamus* (1), *Dendrocalamus* (32), *Dinochloa* (3), *Fargesia* (2), *Gigantochloa* (18), *Indosasa* (4), *Melocalamus* (5), *Phyllostachys* (6), *Pleioblastus* (1), *Pseudostachyum* (1), *Schizostachyum* (3), *Teinostachyum* (1), *Thyrsostachys* (2) and *Yushania* (1) – an astonishing concentration of 94 taxa. About this list several features deserve comment. Firstly, *Chimonocalamus* is a newly described genus. Secondly, one can generalise by saying that, with apparently only nine exceptions, every taxon recognised by the specialist has an equivalent vernacular name. This in itself is remarkable but additionally if one takes, for example, *Dendrocalamus membranaceus*, the taxonomist recognises nine forms in Yunnan within that species (common form, *sulcatus*, *bigemmatus*, *crinitus*, *dimbrili-gulatus*, *pilosus*, *radicatus*, *striatus* and *stigasus*) each of which is found to have a matching vernacular name. A similar situation exists for three variants of *Dendocalamus sinicus*. Finally, in what must be seen as something of a challenge for the taxonomist, of the 18 *Gigantochloa*, for 13 no specific name was attributed. Even so, all but two of these unnamed *Gigantochloa* species have an individual vernacular name. The taxonomists concerned (Wang *et al.* (*ibid.*) remark:

> The indigenous knowledge of cultivation and management of bamboo possesses the rationality, which is often integrated into [the] institution and protective measure[s] for the natural resource utilisation.

It is thought-provoking to ask whether the mental processes of classification in the vernacular and by the specialist are similar since they recognise essentially similar taxa. Wang (1990) commented:

> Their folk classification is not fitted in with the modern taxonomy. They classify plants based on productive practice, social customs, folk legends and economic

uses, habits and shape of plants. They have their own concept of species and ranks above the species. These classifications are successful to some extent, especially in their utilisation. Sometimes taxonomists do not know what to do in the field, but a child may tell him some useful knowledge based on the folk classification. . . . It should be stressed that the bamboo resources of a minority [ethnic] area in southern Yunnan are closely related to food, clothing, shelter and transportation of people. We conclude that the folk classification systems of bamboo are very important.

To a sceptical Western mind this seems a strange way of proceeding and one hardly likely to be adopted. Two reflections however seem apposite. The first is that although Linnaeus attempted an entirely rational approach, both he and ourselves are aware of the notion of 'facies' – obvious resemblances which we perceive ahead of any formal classification and which we cannot altogether describe. Secondly, like that of the Chinese referred to here, our taxonomy is two-fold. With them we share the need, for practical reasons, to know to which plant we refer. Like them, too, we seem to require our taxonomy to have a more intangible quality. In their case it floats off into social custom and folk legend while in our own it takes the form of evolutionary theory and passionately held views about whether this or that arrangement is the more 'natural'. In establishing the eminently practical with more speculative matters it is difficult for us to say how close the two systems are since we have neither practice with, nor close understanding of, the converse approach. It appears however that they do classify, can identify and perpetuate what they know by some kind of consistency in the names they use. It is unclear, however, to what extent they consciously seek to disentangle the more and less practical considerations.

Chorology

Chorology is concerned with the distribution of taxa and seeks to discern informative patterns. Although an investigator concerned with grasses, might have any one of several priorities, some of the interest relates to continental drift.

At the core of modern taxonomy is the familiar hierarchy of family, genera and species and, in the case of grasses, the interposition of subfamily and tribe has proved to be helpful. Self-evidently, a hierarchy handles degrees of similarity. For grasses, closest similarity is among species of the same genus and greatest dissimilarity among those of different subfamilies. As post-Darwinians we make a further assumption, namely that a hierarchy of differences signals the unfolding tapestry of evolution.

If this notion is then superimposed on to that of continental drift, we can proceed in the following way. Finding the Poaceae on all continental land masses, we assume it predated their significant separation, each mass departing with its cargo of 'grass'. If, at the other extreme, species are peculiar to subsidiary parts of individual land masses then their evolution long post-dates fragmentation. And thus the middle levels of the hierarchy come to have special significance. At its simplest, if all subfamilies occur on all land masses then *their* delineation preceded continental fragmentation but if genera are peculiar to this or that large land mass, generic identity (as we perceive it today) is in drift terms late and not early.

On this reckoning the tribe could have a pivotal significance. If tribes are found on all continents delineation at this level was 'early', and if not could it have been 'late'? The joker in the pack is man himself, who as the great traveller disturbs nature's orderly arrangement creating a confusion that his taxonomist descendants must unravel.

Tribal distribution was studied by Hartley (1950) and its detail worked out in a series of subsequent papers (Hartley, 1954,1958a,b, 1961, 1973; Hartley and Slater, 1960), although the hierarchy of the family with which he worked differs in several respects from that currently recognised. None the less, some significant features emerged. His grass groups were arranged with regard to the mean 50°F (10°C) of the midwinter month. Two such temperature boundaries encircle the world, north of the Tropic of Cancer and south of the Tropic of Capricorn, at about latitude 40° in the southern hemisphere and around the same northern latitude but more irregularly. Between these extremes Hartley located a majority of Andropogoneae, Eragrosteae and Paniceae, and outside of these boundaries to the north and south, a majority of Agrosteae, Aveneae and Festuceae. In addition to these climatic preferences what the distribution maps clearly show is all six tribes located on each major land mass. (There is here, clearly, a semantic point. The more fissiparous our view of a tribe, the less likely is it to be found widely distributed.) What we can usefully conclude however from Hartley's evidence is that tribal delineation was well advanced ahead of the wide continental separations, as were the associated climatic preferences, although as the latitudinal differences became more sharply defined so might such preferences also have been sharpened.

It is on the generic level, in practice, that most interest centres. Clayton (1975) remarked:

> ...species distributions tend to be governed by conditions prevailing since the beginning of the Pleistocene, and one instinctively turns to generic distributions for probing further back in time...

Clayton and Renvoize (1986) comment:

> Evidently the genera are not good travellers...

If, therefore, genera are the highest level in the hierarchy restricted to continental land masses then *their* evolution would seem to post-date the major phases of continental separation. This, of course, is something of an oversimplification since the subtribe category can encompass restricted groups of genera. The small subtribes Neurachninae and Spinificinae recognised by Clayton and Renvoize (1986) each encompass genera unique to Australia for example. Given such reservations, the distribution of genera still turns out to be informative.

Clayton (1975) scored the distribution of some 645 genera against 25 geographical areas to reveal not random distribution but significant clustering in groups that came to be referred to as 'phytochoria'. Later work centred on Old World grasses (Clayton and Cope, 1980a) and New World grasses (1980b) and a later résumé of parts of this work is provided in Renvoize *et al.* (1992). A recent contribution is that on Australasian grasses (Cope and Simon, 1995) which includes helpful comment on methodology and interpretation.

Clayton (1975) first recognised seven major groups of genera (from widely different tribal affinities it should be noted), namely Eurasia, North America, Temperate South America, Tropical America, Africa, India/South Seas Asia and Australia. These could then be variously subdivided. Africa, for example, could be separated for this purpose into Africa, South Africa and Madagascar. It was possible then to identify genera shared, for example, between tropical Africa and tropical America.

Later work dealt with newly raised genera and additional distribution data, and two small phytochoria, one from each hemisphere, are presented here in detail from Renvoize *et al.* (1992) to show the diverse taxonomic components (Table 6.2).

What each shows is generic diversity. Even where they share a common tribe they are sufficiently different to be put in separate subtribes where the classification allows. Such examples could be repeated throughout various phytochoria. Those genera with numerous species can be represented in many different situations on a species by species basis.

Poa, with about 500 species, occurs on almost every land mass of any consequence either in the temperate regions or the tropical uplands. *Panicum*, with about 470 species, occurs mostly in warmer regions but in the New World has spread as far north as the Great Lakes.

A puzzle that remains is how to explain the origin and persistence of phytochoria. For different and unrelated taxa to co-exist implies migration into an area of some, at least, of the components. Shared occupancy over time implies either initially or subsequently similar physiological adaptedness but can anything beyond these rather obvious assumptions be added? Clayton (pers. comm.) suggests that phytochoria are primarily determined by climate and coincide with ecological formations, they do not cross wide oceans, are filtered by mountain barriers, can migrate under climatic change and, sometimes, leave traces of having done so.

Table 6.2. Two phytochoria.

(a) Old World: Deccan–Malesian Subkingdom

Genus/species	Subfamily	Tribe	Subtribe
Apluda nautica	PAN	Andropogoneae	Ischaeminae
Arundinella setosa	PAN	Arundinelleae	–
Cymbopogon martinii	PAN	Andropogonae	Andropogoninae
Pseudoraphis spinescens	PAN	Paniceae	Cenchrinae
Saccharum arundinaceum	PAN	Andropogoneae	Saccharinae
Sporobolus diander	CHL	Eragrostideae	Sporobolinae
Thysanolaena maxima	ARU	Arundineae	–

(b) New World: South Tropics Subkingdom (Montane)

Genus/species	Subfamily	Tribe	Subtribe
Calamagrostis effusa	POO	Aveneae	Aveninae
Distichlis humilis	CHL	Eragrostideae	Monanthochloinae
Hordeum comosum	POO	Triticeae	–
Pennisetum chilense	PAN	Paniceae	Cenchrinae
Poa lanuginosa	POO	Poeae	–
Poa ligularis	POO	Poeae	–

Against such a background particular floras can present an interesting challenge, one example being that of Jebel Marra, an extinct volcanic region in the Sudan, studied by Wickens (1976).

Jebel Marra volcanic massif is over 3000 m high and almost equidistant from the Atlantic and Indian Oceans, the Mediterranean and the Red Sea. It is approximately analogous to an island in the sea, the 'sea' being in this case the surrounding and quite different desertified landscape. The mountainous region has a lower temperature and higher rainfall. This in turn creates conditions for a localised flora whose origins have yet to be completely explained. Lower temperatures and higher rainfall in Africa, for example, during the Pleistocene glaciations would have created migration routes for species from southern Europe and extended that for the East African mountain systems. Climatic recession would then isolate such migrants as montane refugia. Later, mankind would bring both intentional and other additions to the flora. Jebel Marra retains long-distance affinity with floras separated from it by many hundreds of kilometres. Only about 30 grass species are common to Tibesti and Jebel Marra which Wickens interprets as showing little Pleistocene migration directly through Tibesti to

Jebel Marra. There is a closer relationship between the floras of Jebel Marra and the East African mountain systems. The story is a complex one eloquently conveyed by some 208 distribution maps of individual species, among them 28 grasses. The Jebel Marra study represents a attempt to explain the coming together of disparate elements in a flora, some natural, some the result of human interference, and provides a meticulously detailed model for probing this kind of problem.

Species Identity

In conclusion, it is worth remarking that the raw material of taxonomic enquiry consists primarily of actual grass specimens taken from the field. They have to be identified, eventually to genus and species, with whatever Flora is available. But field material is variable, prompting questions about how far a specific description is appropriate before a new species is recognised. Such expanding knowledge eventually has implications for our perception of generic limits and higher levels of classification.

This situation explains partly why grass taxonomy is in a state of flux. As our awareness of its diversity increases so our arrangements have to change to take account of it. Even so, present diversity is the outcome of past evolutionary change and this we now examine.

From Extinct to Present-Day Grasses   7

Discovery of the fleshy fruited bamboo in Bahia (Eastern Brazil) supports the hypotheses that the Bahian region harbors relics of ancient grasses... Could there be a relative, as yet undiscovered, of our new bamboo growing along the coast of West Africa? After all, such an area as Cameroon – where some bambusoid grasses are known – was contiguous with the region of Bahia prior to continental drift.

Söderstrom and Calderon (1980)

Issues in palaeoagrostology – the history of the grasses – include the likely time of origin, early evidence of diversification (notably the distinction between bamboo and non-bamboo elements), evolution of C_4 photosynthesis, the supposed co-evolution with herbivores leading to various types of grass-dominated habitat and, much later, the emergence of domesticated cereals.

During the nineteenth century there were attempts to place various grass-like fossils into some sort of order but most of them either belonged to other families or were only doubtfully grasses (Thomasson, 1980). Modern critical study of grass fossils effectively began with Elias (1932, 1934, 1935, 1941 and 1942) who concentrated primarily upon the High Plains of North America. Later workers added to the study of these strata and also explored elsewhere.

Early Records

The relevant part of the geological time scale is shown in Table 7.1 later in this chapter.

Although it might be assumed that (as flowering plants) grasses originated in the Cretaceous, evidence confined to pollen is only available for the ensuing epoch, the Palaeocene (Muller, 1981) lasting some 10 million years. While this establishes the presence of the Poaceae, the information is of limited use since across the family pollen is remarkably uniform even in electron microscope surface detail (Page, 1978).

Eocene

In 1964 Chandler presented evidence for four kinds of caryopsis from southern England. Thomasson (1987) critically reviewed evidence for grass fossils and of his three categories 'undoubted', 'probable' and 'possible' only one record, *Graminophyllum*, fell into the first category for the Eocene.

Attempts to resolve the near absence of grass fossils, when they might have been expected, have involved recourse to 'phytoliths' ('silica bodies', 'plant opalines') that form in epidermal cells in shapes and characteristic of different grasses (Prat, 1932). These can persist in soil almost indefinitely but lack sufficient diversity to allow a particular genus or even tribe to be identified unmistakably. Even allowing Thomasson's second category 'probable' did not materially change the situation. Then, in 1991 Crepet and Feldman illustrated a fossil spikelet with two florets, an SEM of pollen from the spikelet and a rhizome with leaf sheaths from the Palaeo/Eocene Wilcox Formation in western Tennessee. The material appeared to have affinities with either Pooideae or Arundinoideae. A year later in 1992 Poinar and Columbus presented the remarkable photographs of *Pharus* (Bambusoideae) fruit attached to a mammal hair all preserved in Dominican amber. Assuming the dating of this and the previous example to be correct, grass diversity was probably significant by the end of the Eocene although, based on their modern counterparts, the grasses might have been uniformly C_3.

Amber is believed to have been preserved from at least Cretaceous times (Poinar and Columbus, *ibid.*) and it may yet reveal more instances relevant to the study of grasses.

Oligocene

Again from North America, the Florissant Beds, reckoned as early Oligocene or perhaps late Eocene, have yielded grass fossils (MacGinitie, 1953; Beetle, 1958; Epis and Chapin, 1975). What makes this of particular interest is that the grasses there are not only pooid following Clayton and Renvoize (1986) but belong to separate tribes. The genera are tentatively attributed as *Stipa* (Stipeae) and *Phalaris* (Aveneae). If one accepts the view

Table 7.1. Evolution of the horse and significant aspects of grass biology within the geological time scale.

Period	Epoch	Duration (Ma)	Start of time period (Ma)	Evolution of the horse[a]	CO_2 level[b] ($\mu l\ l^{-1}$) approx.	A sample of items and events relevant to Poaceae
Quaternary	Holocene	Approx. last 10,000 years				Origin and development of agriculture. Withdrawal of the ice sheets
	Pleistocene	2.5	2.5	*Equus* (1 large toe only)	250	Replacement in N America of stipoid grasses with *Andropogon, Bouteloua, Buchloë* and *Panicum*. Migration to S America of stipoids. Panamanian land bridge. Migration of ungulates from N to S America (Stebbins, 1981). Reciprocal migration across Bering Bridge (Simpson, 1951)
Tertiary	Pliocene	4.5	7	*Neohipparion, Hipparion, Nanippus* (all three toed) *Pliohippus* (one toed)	250	Expansion of grassland, contraction of savanna and woodland in western United States (Shotwell, 1961)
	Miocene	19	26	*Merychippus* (hypsodont, one large, three small toes)	250	Expansion of C_4 ecosystems as an indicator of global change (Cerling *et al.*, 1993). Oldest identifiable fossils with bundle sheath layers (Thomasson *et al.*, 1986). *Stipa anthoecia* (Elias, 1932). Main adaptative radiation of grass tribes (?) (Campbell, 1985). Chloridoid and panicoid fossils identified from E Africa (Retallack *et al.*, 1990)

Table 7.1. (contd)

Period	Epoch	Duration (Ma)	Start of time period (Ma)	Evolution of the horse[a]	CO_2 level[b] ($\mu l\ l^{-1}$ approx.)	A sample of items and events relevant to Poaceae
Tertiary (contd)	Oligocene	12	38	Mesohippus (three toed)	250	Late Oligocene global aridisation (Zubakov and Borzenkova, 1990). Mammals acquired hyposodont teeth (Stebbins, 1981). Phalaris and Piptochaetium in Florissant Beds (Beetle, 1958). Later estimate for emergence of savanna (Clayton, 1981). Grass pollen abundant from now on (Muller, 1981)
	Eocene	16	54	Hyracotherium (Eohippus)	500	Caryopses (Chandler, 1964). Pharus (Bambusiodeae) spikelet in amber adhering to mammalian hair (Poinar and Columbus, 1992). Grass pollen frequent (Muller, 1981). Earlier estimate for emergence of savanna (Stebbins, 1987)
	Palaeocene	11	65	Primitive ungulates (condylarths), browsing with low-crowned teeth and five toes. Widespread	750	
					250	Firm reports of grass pollen (Muller, 1981)
	Cretaceous	71	136		2000–3000	Doubtful reports of grass pollen (Muller, 1981)

[a] After MacFadden (1994).
[b] After Ehleringer et al. (1991).

of Watson and Dallwitz placing the Stipeae among the arundinoids then these two fossil grasses can seem more divergent.

In late Oligocene/early Miocene times there appears to have been a trend toward drier climates.

Miocene–Pliocene

An area of the central United States (Nebraska, Kansas, Colorado and New Mexico) contains a fossil sequence first explored in detail by Elias (1932, 1934, 1935, 1941 and 1942) of which the essentials, for our purpose, are as follows. In early to mid-Miocene a genus *Berriochloa* was found that subsequently became extinct. By the lower (early) Pliocene *Nassella*, *Oryzopsis*, *Piptochaetium* and *Stipa* were present in well-differentiated forms, all it should be noted C_3 as judged by their modern counterparts. In passing it should be noted too that these grasses are different genera to those which today occupy the same region, a point taken up later in this chapter.

From apparently late Miocene, Thomasson *et al.* (1986) found evidence of a 'chloridoid' fossil leaf anatomy. With one apparent exception (*Eragrostis walteri*) all chloridoids are believed to be C_4, suggesting this type of photosynthesis was functioning about 7–5 million years ago. Co-existent with this find were fossils of the three-toed horse *Neohipparion* discussed later.

Retallack *et al.* (1990), studying mid-Miocene soils at Fort Ternan in south-west Kenya, described three panicoid and two chloridoid grass fossils based on stomatal morphology and phytoliths. Given the dating of this material and assuming at least the chloridoids to be C_4, this type of photosynthesis would have been operating 14 million years ago. This would imply the existence of C_4 grassland perhaps 5 million years before the divergence of humans (adapted to more open habitats) from apes (more forest dwellers). Other evidence however points differently.

C_3 plants discriminate against $^{13}CO_2$ relative to $^{12}CO_2$ but C_4 plants do not. Organic matter of C_4 plants therefore has a $^{13}C:^{12}C$ ratio similar to that of the atmosphere while that of C_3 plants is relatively depleted of ^{13}C. $\delta^{13}C$ values for C_3 plants range from -35 to $-22‰$ while those for C_4 plants are from about -18 to $-9‰$.

For fossil soils at Fort Ternan, Cerling and Quade (1991) found $\delta^{13}C$ values of around $-27‰$ implying a strongly C_3-dominated vegetation – perhaps closed or nearly closed canopy forest.

How might such contrary evidence be reconciled? The area was prone to ecological disturbance due to volcanic ash and transgressive and regressive lake events. Cerling and Quade (*ibid.*) point out that fossil soils formed over thousands of years and their $\delta^{13}C$ values reflect a long-term situation. Perhaps therefore the presence of C_4 grasses at Fort Ternan represents a transitory situation, the colonising of a disturbed habitat prior to re-establishment of the forest canopy. Whatever the explanation, the fossil chloridoid material, if authentic, substantially shifts forward in time the existence of C_4 photosynthesis.

Another source of fossil data arises from exploratory oil drillings. One such example is by Morley and Richards (1993) from the Niger Delta in West Africa. Material from savanna fires is assumed to have been deposited after long-distance travel in the atmosphere in the Niger Delta. Cores from mid-Miocene show charred grass cuticle suggesting savanna fires far more frequent than in the early Pliocene. These authors did not distinguish between C_3 and C_4 grasses.

After the Pliocene

Our assumption is that from the Pleistocene onwards a 'modern' grass flora existed substantially similar to that of the present day though more regionally confined. Before the advent of humankind there were fewer opportunities for long-distance transport of grass propagules.

Recent

In the Near East carbonised remains of cereals are known from about 8000 BC (Helbaek, 1959). In Central America 'prehistoric' maize remains preserved by natural dehydration have provided insights into the origins of its domestication. These materials are considered in more detail subsequently.

The Spread of Grasses

Although chloridoid (presumably C_4) grasses date back to perhaps mid-Miocene, on present evidence uncertainties remain. Not only is the fossil record, globally, extremely fragmentary, but we also need to gauge the effects of climatic change. It is theoretically possible that the outlines of the five major subfamilies were in place by the Eocene. Our failure to detect them might then be due to either absence of fossilation, or an adequate fossil record not yet sufficiently explored or, for climatic reasons, these early grasses having had only limited distribution amid high canopy woody vegetation. Perhaps only significant aridification of the climate in late Oligocene times created favourable conditions for grasses to be widespread. The alternative possibility is that C_4 photosynthesis was originated in (say) the early Miocene and that even our fragmentary fossil record does not mislead us on this point.

C_4 and atmospheric CO_2

Table 7.1 includes estimates of CO_2 levels from the Cretaceous onward. If one accepts that C_4 photosynthesis 'economises' CO_2 available to the plant, the lowest values, and therefore the greatest relevance of C_4 photosynthesis, occurred in the Palaeocene and the Miocene. If additionally, significant aridification occurred in the late Oligocene–early Miocene C_4 photosynthesis could have originated in the Palaeocene though Miocene is perhaps more likely. The matter is considered in some detail in the next chapter.

Co-evolution

Biology has many examples of co-evolution. These include the co-evolution of figs and fig wasps each heavily dependent on the other, but *mutual* dependence need not be complete. While maize, for example, depends on humankind for its survival, we are well able to exclude it altogether from our diet. On the world's grasslands herbivores have evolved. To what extent is the grass–herbivore relationship one of 'in step' co-evolution? Alternatively, is the arrangement relatively one-sided with one partner only being heavily dependent? To explore these possibilities, we examine the matter utilising relatives of the horse. The taxonomic arrangement here follows MacFadden (1994). Among surviving relatives in the Order Perissodactyla are tapirs, horses and rhinoceroses. Horses, the Equidae, form a family within which are a series of subfamilies. These are Hyracotheriinae, including *Hyracotherium*, and Anchitheriinae, including *Mesohippus* and *Parahippus*. Both these subfamilies are now entirely extinct. The third subfamily Equinae contains two tribes, Hipparionini which includes the extinct genera *Merychippus* and *Hipparion*, and Equini containing the extinct *Pliohippus* and *Equus* the sole living genus. *Equus*, or the horses, are of worldwide distribution. They exist as several species, namely *E. caballus caballus* (domestic horse), *E. caballus przewalskii* (the wild Przewalskii horse), *E. hemionus* (Asiatic onager or wild ass), *E. africanus* (African ass) and *E. burchellii*, *E. zebra* and *E. grevyi*, the last three being different kinds of zebra.

In examining co-evolution of grasses and herbivores, the horse is a convenient example since its fossil lineage is so thoroughly documented. Table 7.1 includes seven of the 33 extinct genera of the horse currently recognised by MacFadden (*ibid.*) in their geological epochs.

Evolution of the horse summarised

During the Palaeocene a five-toed mammal group, the Condylarths, was widespread. Their nearest modern equivalents are tapirs – browsing animals with four-toed front and three-toed rear feet.

In the Eocene, *Hyracotherium*, more familiar as 'Eohippus', appeared on the continental land masses that became the New and Old Worlds. It was between 25 and 50 cm high at the shoulder with low-crowned teeth and with four toes on the front and three on the rear feet. (By reference to the human hand this is equivalent to having lost the little finger on the front and this and the thumb on the rear feet. Our long middle finger corresponds to what became the hoof and the ring and index fingers the 'side toes' and the thumb became progressively diminutive.)

During the Oligocene a larger three-toed horse, *Mesohippus*, evolved, which stood about 60 cm at the shoulder. Judged from the skull interior, the brain was larger and more convoluted than that of *Hyracotherium*, indicating improved mental capability. This forerunner of the horse had low-crowned teeth and would have browsed rather than grazed. *Mesohippus* graded into *Miohippus* surviving into the Miocene. Assuming grasses to have been available by the beginning of the Palaeocene, this represents co-existence with the browsing horse ancestors for about 40 million years to the beginning of the Miocene. During the Miocene several major changes occurred implying a closer interaction between the 'protohorses' and grasses, which can be summarised as follows:

1. *Miohippus* survived into the Miocene.
2. There then radiated four groups, *Parahippus*, *Archaeohippus*, *Hypohippus* and *Anchitherium*, all with relatively low-crowned teeth (though higher than in *Miohippus*) committed to browsing and all with three-toed feet.
3. From the first of these, *Parahippus* or something closely akin, there arose *Merychippus* whose teeth possessed 'cement' making them better adapted to grinding a coarse plant diet.
4. Increasing aridity both widened the area available for grasses and created the circumstances for their diversification. (The possible origin of C_4 photosynthesis around this time was mentioned above.)

It is important to note just what is being claimed here. The spread of grasses is seen primarily as a response by an opportunist plant family to climatic change. Given such circumstances, a disadvantaged tree flora might suffer under browsing pressure but this would be more likely a contributory rather than a primary cause of its decline. The existence of a more open, grass-dominated environment appears then to have created an opportunity for animals with higher-crowned teeth to graze such tough siliceous plants and, with some degree of speed, to escape potential predators.

Much of the foregoing derives from Simpson (1951). More recently MacFadden (1994) has reconsidered hypsodonty in this connection. Grazers ingest not only forage but, as dust, the substrate upon which the grasses are growing. MacFadden (*ibid.*) has pointed out that fine-grained Oligocene clays were relatively widespread but that during the late Oligocene–early Miocene, there was a change over to coarser-grained sands. The transition coincided with the evolution of hypsodonty among horses and could therefore be a contributory factor. MacFadden also points out that not all hypsodonts are grazers, for example the gazelles and certain grazers such as baboons and kangaroos are not hyposodont.

Integral to this discussion is the notion that a three-toed hoof is adapted both to running forwards and dodging sideways. A hoof with smaller side toes would be useful in open country but not well adapted to dodging. Indeed, under these conditions side toes which merely projected but did not function might be more prone to injury. In summary, a three-toed hoof was adapted to closed or nearly closed canopy conditions while the single-toed hoof is appropriate to clear stretches of open country.

Perhaps then, it is not surprising that after the Miocene conditions we find among the descendants of *Merychippus* the three-toed *Hipparion*, *Neohipparion* and *Nanippus* and the first one-toed horse, *Pliohippus*, the direct ancestor of *Equus* the modern horse. Shotwell (1961) for the Great Basin region of the western United States has compared the distribution of the divergent *Merychippus* lines that led to three- and single-toed derivatives. With the spread of grassland through the Pliocene, *Hipparion* and *Neohipparion* became more confined as woodland and forest diminished. The expansion of grassland was accompanied by increasing evidence of *Pliohippus*.

At the end of the Tertiary period in the changeover from Pliocene to Pleistocene it is possible to identify a series of events that impinge on the co-evolution of grasses and horses.

1. A land bridge formed across the Panama region linking North and South America.
2. The descendants of *Pliohippus* and other animals migrated across it.
3. It opened up a migration route for plants, including grasses.
4. Grasses not found in the South American fossil record (*Nassella*, *Oryzopsis*, *Piptochaetium* and *Stipa*) might around this time have entered from the North.
5. During the Pleistocene, the modern horse *Equus* crossed not only into South America but also via the Bering Bridge into Asia and thereafter spread and diversified in the Old World.
6. In a curious and unexplained way horses died out throughout North and South America. Although North America was the site of horse evolution it is now repopulated, through human agency, with descendants of survivors that had earlier migrated to the Old World.

Ashfall Fossil Beds rhinoceros Teloceras major

Although the foregoing discussion has, for convenience, centred upon the horses, another member of the Perissodactyla deserves comment. This is *Teloceras major*, an extinct rhinoceros from the late Miocene (MacFadden, 1994). Specimens, which included some 200 virtually complete skeletons, were remarkably preserved in what was Poison Ivy quarry, now Ashfall Fossil Beds State Historical Park, a truly extraordinary site in north-eastern Nebraska. From foodstuffs preserved in the teeth and presumed digestive tract, Voorhies and Thomasson (1979) showed that the now extinct grass genus *Berriochloa* was conspicuous in the predominantly grass diet.

Features of North American Grasslands

During Pleistocene/Holocene times, the grass flora of North America has changed. Stipoid grasses that accompanied the evolution of the horse have been replaced by bluestems (*Andropogon* spp.), switchgrass (*Panicum virgatum*), grama grass (*Bouteloua* spp.) and buffalo grass (*Buchloë dactyloides*) that migrated northward with increasing continentality of the climate. Further north *Agropyron* and *Elymus* are more common.

During the Pleistocene, bison and sheep arrived from Europe via the Bering Bridge. Since they graze more closely than do horses, they would favour these more recently arrived grasses that form a turf, rather than the clump-forming stipoids (Stebbins, 1981).

Bamboo Herbivores

Perhaps the strangest examples of herbivory involve three oriental mammals: *Ailuropoda melanoleuca*, giant panda, *Ailurus fulgens*, red panda, and *Rhizomys*, a genus of three species, the bamboo rats. All of these animals have a year-round dependence on bamboos, the giant panda consuming stems and leaves, the red panda mostly leaves, and the rats the underground parts. The following outlines giant panda dependence on bamboo.

Ailuropoda is, surprisingly, by reference to dentition and digestive apparatus, a carnivore having a relatively short digestive tract and no ruminant-type stomach nor enlarged caecum. The plant to which it is so committed is relatively both nutritionally poor and high in silica. To make the best therefore of this unpromising liaison the panda must feed almost continuously and yet minimise demand on its energy reserves. At Wolong reserve where giant panda diet has been most closely studied there are seven species of bamboo: *Fargesia robusta* (umbrella bamboo), *Phyllostachys heteroclada*, *P. nidularia*, *Sinarundinaria chungii*, *S. confusa*, *S. fangiana* (arrow bamboo) and *S. ferax*. Arrow and umbrella bamboos are those important to the panda.

Bamboos have persisted from the Eocene as indicated by *Pharus*, preserved in amber (Poinar and Columbus, 1992) for some 40 million years, and traces of panda lineage first crop up in the fossil record some 20 million years ago in the Miocene (Schaller, 1993). The giant panda appeared in late Pliocene or early Pleistocene times about two or three million years ago.

At some point, whether the giant panda emerged from either bear or racoon lineage, it shambled off into the uncertain future of a diet dominated by bamboo. To cope with this situation, natural selection seems to have provided the panda with premolars adapted for grinding, a massive jaw and associated muscles, an especially tough stomach and a curious 'pseudo thumb' on each front paw formed from a modified metacarpal bone. The impression from seeing pandas at close quarters is of a stolid paradox – an herbivorous carnivore physically very powerful but, for all that, desperately vulnerable to both the diminution of its habitat and the flowering behaviour of bamboo. Such are the dietary preferences of the giant panda that it is dependent not on just any bamboo but especially on two species prone to gregarious or mast flowering. Laidler and Laidler (1992) instance a population of 150 giant pandas being reduced to 20 over 11 years through bamboo die-back following flowering at Wolong. This kind of result raised awareness of the panda's precarious status and led eventually to the sombre realisation that constriction of its habitat through human exploitation was a severe contributory problem.

The red pandas, though dependent on bamboos, currently have larger populations, and, although relatively rare, are hardly as endangered as is the giant panda. The bamboo rat is not at risk.

Bamboo rehabilitation

The situation regarding the giant panda has stimulated interest among botanists in the bamboos upon which the animals depend. The aims have been to understand better the mechanism of flowering and to rehabilitate the natural habitat.

Taylor (1988) reported the results of a detailed study of regeneration from seed of *Sinarundinaria (Bashania) fangiana* in the Wolong panda reserve. This bamboo in 1983 flowered, set seed and died over 80–90% of its range, a situation in which pandas either move to a different unflowering species, seek unflowering patches of the preferred species, perhaps emigrating some distance, or simply starve.

Sporadic flowering had begun about 1970, increased spasmodically and reached a climax in 1983. Even after this, small unflowered patches remained especially at higher altitudes around 3000 m.

Experimentally, seed germination and establishment were studied in 90 plots that included, with and without shade and similarly for fertilisers. Both by locations and year of establishment recruitment was greater in 1985 than in 1984. Selective felling was more conducive to establishment than clear felling.

Reid *et al.* (1991) recognised the value of selective rather than clear-cut felling in aiding the establishment of *Sinarundinaria fangiana* seedlings. Taylor *et al.* (1991) recommended establishment of this bamboo species under conifer, *Abies faxoniana*, for preference, pointing out that mass flowering be seen as an opportunity to restore panda habitats and that a similar one, on present expectations of inter-mast interval, would not present itself until about AD 2028.

Although the developed world might wish to give the panda high priority this is not necessarily the case with peasant farmers seeking a livelihood in or near its habitat. Resettlement, though possible, is not straightforward and in practice farmers and their families normally resist pressure to be moved onto new holdings.

Many interests, zoological, botanical, political and sociological among them, converge on the well-being of the panda. For botanists, improved husbandry of arrow and umbrella bamboo is clearly important.

The enigma of panda diet remains. As Laidler and Laidler (1992) point out:

> Schaller and his co-workers report that a giant panda will walk quickly through stands of nutritious wild parsnip (a favourite bear food) to feed on a nearby clump of much lower quality bamboo. Why, when the giant panda is balanced on a knife edge of nutritional deficit which, impinges on every aspect of its behaviour, does the creature literally walk past food that could make its life a lot less difficult?

The Grass–Herbivore Relationship: An Assessment

In the Eocene long before the evolution of the horse, grass epidermal cells contained phytoliths. Browsing animals with low-crowned teeth doubtless occasionally browsed tall grasses, but presumably such grasses can hardly have been a preferred item of diet. The evolution of the three-toed horses coincided with the diminution of tree cover and the development of a savanna type of habitat – tall, clump-forming grasses under scattered tree cover. This gave way to more open grassland within which one-toed horses were at an advantage. Almost certainly the changes in the flora were primarily climate-driven favouring one herbivore rather than another. The phytoliths, long the possession of grasses, persisted and a consequence of changes in horse dentition was improved ability to cope with an increasingly grass-based diet.

With the demise of the horse in the New World, it is worth remembering that the immense herds of bison that roamed North America when the first European settlers arrived were themselves immigrants from the Old World. Likewise, the C_4 grasses upon which they grazed were immigrants from Central America (e.g. *Andropogon, Panicum, Bouteloua* and *Buchloë*). Grasses were the primary opportunists, secondarily exploited by animal opportunists.

How much grasses owe to herbivores is less certain. Grass fruits can sometimes survive passage through the animal gut, they can travel, for example, on horse fetlocks, and even before these herbivores evolved *Pharus* could attach to a mammal hair. Anyone who has observed the difference between (say) sheep-grazed and ungrazed *Festuca ovina* is aware of grass 'plasticity'. It is unclear how much of this plasticity was in place as part of the grass constitution before the advent of grazing herbivores. From what is known of *F. ovina* (see Chapter 9) its present features include seemingly slow rates of change and long-term occupancy of many of its habitats.

Those animal activities which make for open habitats, grazing and trampling, create opportunities for grass colonisation. Spiky appurtenances to the caryopsis can, to some extent, with or without endophytic fungi, impede grazing and so allow seed maturation. There is clearly some dependence of grasses on animals, but less so in low rainfall regions that would in any case tend to maintain open habitats unaided.

A note of caution is appropriate. Grasses can pose serious difficulties for grazing stock. Although *Stipa capensis* for example, as shown in Chapter 2, is a component of low-grade pasture that sheep will graze, it should be avoided when in flower. The awns are sharp and can enter the eyes, mouth and other tender areas, causing the sheep to have inflammation and sores and damaging the quality of the fleece.

The Southern Hemisphere

Geological events in the southern hemisphere have left behind consequences for plant life.

The woody dicotyledon genus *Hebe* occurs in Chile, Tasmania, New Guinea and New Zealand. Distribution of the southern beech, *Nothofagus*, is similar. A gymnosperm *Podocarpus* also occurs in Chile, Tasmania and New Zealand apart from elsewhere.

Such 'disjunct' distributions, of which there are other examples (Good, 1964), are explained by assuming that they are surviving outliers of an

immense forest that covered Antarctica – a conclusion already accepted by
Darwin in 1859. Two considerations should temper our surprise. Firstly,
for much of its existence Antarctica (as we term it) was less southerly and,
secondly, the cold of the polar regions has not always been so extreme. This
view is supported by drill cores, for example from Seymour Island, off the
tip of Grahamland, the conspicuous Antarctic peninsula pointing toward
Tierra del Fuego. Such cores from the Tertiary yielded pollen of
Nothofagus, *Podocarpus* and other genera (Cranwell, 1959). A point
of interest was the absence of grass pollen, suggesting a dense forest
canopy.

Much further north, a warm-climate family, the Proteaceae, is shared
by South Africa and Australia. *Protea* occurs in the former and the distinct,
though recognisably related, genera *Banksia* and *Grevillea* occur in the
latter.

Do these thought-provoking distributions have equivalents among the
grasses? Before attempting to answer this, we examine the drifting southern
continents more closely.

Moving continents

If one regards somewhere around the southern tip of Latin America as a
pivot, a huge land mass comprising Antarctica, Australia, New Guinea and
New Zealand and India detached and subsequently fragmented further.
Antarctica was flung into what became a polar deep freeze. Australasia
(including New Guinea) travelled east rotating slightly anticlockwise as it
did so. Subsequently New Guinea detached northward and New Zealand
to the south-east. Australia became more arid, New Guinea, tropical, and
New Zealand, a much smaller land mass than Australia and at a cooler
latitude, developed a moist oceanic climate, a situation somewhat akin to
Tasmania (Table 7.2 based on Wilford and Brown (1994)). The changing
climate during the Tertiary of Australia has been lucidly described by
Trusswell (1993). Although there were low frequencies of grass pollen
found back as far as the Oligocene, only toward the end of the Tertiary and
into the Quaternary period is there convincing evidence of a widespread
Australian grass flora (Truswell, pers. comm.). Around Antarctica and in
the southern Atlantic and the southern Indian Oceans are islands, some left
by migrating land masses, some of volcanic origin. The floras of these as
well as those of the larger land areas, provide pieces of a jigsaw puzzle that
changes through time. How much each of these islands preserves under
oceanic conditions remnants of the Antarctic flora and how much these
new insular environments are the sites of adaptation and change are
questions of which we must be aware. Table 7.3 indicates the nature of the
circumpolar grass flora.

On this basis, Brazil, having occupied its position across the equator since the Carboniferous about 300 million years ago, represents relative stability while other areas as they swung across the latitudes provided more adaptive challenges for the plants inhabiting them. This reached an extreme in Antarctica. Here vegetation is at its limit and maintains only tenuous occupancy at the margin of this hostile glacial desert. One component here is the grass *Deschampsia antarctica*.

Against this background, seven grasses will be examined.

1. *Anomochloa marantoidea* (Bambusoideae, Bahia province, Brazil)

Judziewicz and Söderstrom (1989) detail how this species was first discovered about 1842, and described in 1850 and 1851, having been grown first in Paris and then Kew. Subsequently, in the wild, it was lost to science until, after deliberate search, it was rediscovered in 1976. It has $2n = 36$ chromosomes (Hunziker *et al.*, 1989).

Table 7.2. Moving continents in the southern hemisphere (based largely on Wilford and Brown, 1994).

Ma	Significant events
	First human settlement of New Zealand probably within the last 1000 years
	Periodic expansions of polar ice sheets lowered sea level making land connections from Australia to New Guinea and Tasmania
10	Continued expansion of Antarctic ice sheet caused aridity in Australia
20	Northern fringe of Australia bordering tropical monsoon climate. Sri Lanka crossing the equator
30	Ice sheet established in East Antarctica. Major mountain building in New Guinea
40	Deep marine strait established between Tasmania and Antarctica initiating oceanic movement around the polar continent and with increased cooling there
50	Uplift of mountains in Antarctica further encouraging snow and ice there
60	Australia continuing to drift away from Antarctica. Ice building up in Antarctica
70	India continuing to drift northwards, its northern extremity reaching the equator
80	Madagascar separates from the Indian land mass
90	Isolation of New Zealand. Increasing separation of New Guinea from Australia
100	The northern extremity of Antarctica had moved southwards to about the Antarctic Circle
110	Opening up of Tasman Sea separating Australia and New Zealand
120	Substantial seaways had opened between Greater India and Australia and between Australia and Antarctica
130	Southern Atlantic began to open proceeding northwards. The southern extremity of Antarctica has crossed the South Pole
140	Separation of India, Australia and Antarctica becoming apparent
150	Movements among Africa and South America, Antarctica and Greater India and Australia had been initiated

It is confined to a small area and, as its name implies, it is anomalous. Its habitat is threatened and material raised elsewhere had seemed to lack the vigour of the native-grown plant. It is not equipped for long-distance seed dispersal. Despite its importance for grass science it seemed extremely vulnerable and might well have become extinct. An assumption made elsewhere in this book is that bamboos represent our most primitive surviving grasses or at least give us good indications from their floral structure of how earlier more primitive grasses might have appeared. The bamboos, as was pointed out in Chapter 4, are themselves far from uniform. It may be that we shall come to recognize that not only do *Anomochloa*, *Streptochaeta* and *Streptogyna* differ widely from each other, but additionally we might need to see them as other than 'bambusoid'. *Anomochloa*, through the efforts of American botanists and horticulturists is now being more widely grown and under careful management. Latterly therefore, although vulnerable in its native habitat, its prospects both of survival and of being available to botanists for detailed study have markedly improved (L. Clarke, pers. comm.). Quite apart from theoretical considerations, such practical matters underline our prospects for a better understanding of grass origins. At present, the situation for another rare grass *Glaziophyton* is less encouraging.

The obvious assumption hitherto was that *Anomochloa* had survived in a largely unchanging habitat, to which it is well adapted, for millions of years. It is a small specialised genetic construct almost certainly unable to hybridise and with little prospect of release from relict status apart from human intervention.

2.   Streptochaeta spp. (Bambusoideae, New World tropics)

Anomochloa is regarded as 'strange'. *Streptochaeta* is considered 'primitive', perhaps the most primitive genus in the Poaceae. Three species are recognised: *S. sidoroana* $2n = 22$ (Mexico, Panama, Ecuador and Peru), *S. spicata* $2n = 22$ (Mexico, Trinidad, Paraguay and Brazil) and *S. angustifolia* (Brazil) (Judziewicz and Söderstrom, 1989).

At maturity, the spikelets dangling by their coiled awns can become entangled with animal fur and thus be dispersed. It is tempting to suppose that living in an isolated, climatically fairly uniform part of South America it was enabled to move north, with the rise of the Panamanian land bridge, to an approximately similar environment in central America.

3.   Cortaderia (Arundinoideae, southern hemisphere)

Of this genus there are perhaps 25 species with 19 in South America, four in New Zealand and one in New Guinea. Neglecting Antarctica, this grass is thus at the periphery of the area under discussion.

Renvoize and Clayton (1992) speak of the subfamily Arundinoideae as having 'retreated' to the southern hemisphere. Of *Cortaderia*, Clayton and Renvoize (1986) mention gynodoecism (female and bisexual plants) and indicate the adoption of apomixis. This is described for *C. jubata* by Philipson (1978).

The disjunct distribution of this genus is consistent with the idea of drifting continents.

Even if one were disposed to regard *Cortaderia* as a relict genus in a declining subfamily, *C. selloana* (Pampas grass) flourishes around the world under human influence as a vigorous and impressive ornamental.

4.   *Alloteropsis* (Panicoideae, Old World tropics)

From five to eight species are recognised distributed throughout Africa, Asia and Australia. Of these, the most interesting is *A. semi-alata*, perhaps more usefully regarded as a 'species complex' and found widely distributed.

Throughout Africa, Madagascar, India, South East Asia, the Pacific islands and Australia, *A. semi-alata* subsp. *semi-alata* is C_4. *A. semi-alata* subsp. *eckloniana* is C_3 and confined to southern Africa (Gibbs Russell, 1983). For *A. semi-alata* intermediate C_3–C_4 types with $\delta^{13}C$ values between -11.8 and -26.6% have been found in tropical East Africa (Hattersley and Watson, 1992).

It is not possible to say how far the present-day distribution of *A. semi-alata* subsp. *semi-alata* has been extended by human activity but it is noteworthy that areas once linked continentally share C_4 photosynthesis. There is, however, an interesting complication considered in the next chapter in that in Australia it is PCK (Prendergast *et al.*, 1987) and in South Africa NAD-ME (Frean *et al.*, 1983).

5.   *Tripogon* (Chloridoideae, mostly Old World tropics)

Of 30 species, one only occurs in South America. *Tripogon* typically grows on wet flushes on and around rock outcrops but in two regions, southern Africa and Australia, has evolved 'resurrection' types *T. minimus* (illustrated in Chapter 2) and *T. loliiformis* respectively. Such grasses can tolerate severe dehydration but revive rapidly with the onset of rain (Gaff and Ellis, 1974; Gaff and Latz, 1978; Lazarides, 1992).

Other resurrection grasses are known (*Eragrostiella bifaria, Eragrostis nindinensis, Oropetium capensis* and *Sporobolus portobolus* (among the chloridoids), *Micraira* spp. (arundinoid) and *Poa bulbosa* (pooid)) but it is surely significant that chloridoids are prominent in this respect given the affinity this subfamily has for arid and semi-arid environments. With prolonged exposure to arid conditions on separate continental land masses it is from the chloridoids that most resurrection grasses have arisen. Such grasses (in any subfamily) have not been reported from South America but it is unclear whether they do not grow there or have not yet been the object of thorough search.

The suggestion here is not of already evolved resurrection grasses being dispersed to South Africa and Australia but arising in each (from mostly chloridoid) grasses as desertified environments developed probably at appreciably different times.

6. Deschampsia antarctica

Whatever may have grown on the Antarctic mainland in times past it is at present almost completely inimical to flowering plants. Edwards and Lewis Smith (1988) note *Deschampsia antarctica* as the only grass and one of only two species of flowering plant that occur there. (The other is a pearlwort *Colobanthus quitensis* in the family Caryophyllaceae.)

Given the Ice Age history of Antarctica it is almost inconceivable that *D. antarctica* would, since Tertiary times, have maintained uninterrupted occupancy of the mainland. The ameliorating effects of the sea on the climates of the circumpolar islands make it likely that *D. antarctica* would, probably through the agency of birds, be periodically reintroduced from such locations into mainland Antarctica.

D. antarctica can photosynthesise at 0°C at about 30% of its maximum. Edwards and Lewis Smith (*ibid.*) comment that *D. antarctica*, while tolerating salt spray, does not seem, among arctic and alpine species, to have developed any unique metabolic adaptations for survival under Antarctic conditions.

7. Poa flabellata

In Table 7.3 this species is reported from South Georgia and Macquarrie but grows on other islands too. It was formerly more common in the Falkland Islands, both the principal land masses there, and various smaller islands of the archipelago. Through burning and unrestricted grazing the incidence of *P. flabellata* has declined except in some protected areas, particularly in the smaller islands. It is tussock forming and is locally called 'tussac'.

Compared with the more familiar *Poa* species such as *P. annua* and *P. pratensis*, this grass has a remarkable growth habit. Woods (1970), an ornithologist, writes:

> Tussac can grow to 3.5 m (11½ feet), forms dense monospecific stands and is very long-lived. Smith and Prince (1985) obtained a radio carbon date from the pedestal of a large healthy plant on Beauchêne Island of about 300 years. A mature tussac plant produces a fibrous pedestal (tussac bog) 1–1.5 m (3–5 feet) high from the crown of which spring the flower stems and thousands of long leaves that remain green for about two years. Foliage that has died eventually folds down, remains attached to the base and forms a skirt projecting round the pedestal. Tussac leaves are tough but, fortunately, do not have sharp or serrated edges. Nevertheless, adjacent plants in mature tussac grassland are so close that the skirts interlace completely and are inseparable.

Woods goes on to describe how the heavily shaded ground beneath provides a habit for various burrowing birds notably shearwaters and diving petrels and whose excreta fertilise the tussock-grass stands. Woods quotes for example Kidney Island, only 32 ha in area but well covered with *Poa flabellata*. He estimated 5000 breeding pairs of birds of some 28 species.

Table 7.3. The circumpolar grass flora (based on Greene and Walton, 1975).

Location	Approx. area (km^2)	Lat. °S	Genera	Species
Main Antarctic land mass	14,245,000	90–66	Indigenous	
			Deschampsia	antarctica
			Alien Poa	annua, pratensis
Kerguelen	6200	49	Indigenous	
			Agrostis	magellanica
			Deschampsia	antarctica
			Festuca	contracta
			Poa	cookii, kerguelensis
			Alien Agrostis	canina, stolonifera
			Anthoxanthum	odoratum
			Arrhenatherum	elatius
			Holcus	lanatus
			Poa	annua, pratensis, trivialis, etc.
Macquarrie	123	55	Indigenous	
			Agrostis	magellanica
			Deschampsia	chapmani, penicillata
			Festuca	contracta
			Poa	foliosa, hamiltoni
			Puccinellia	macquariensis
			Alien Avena	fatua
			Poa	annua
Prince Edward	47	46	Indigenous	
			Agrostis	magellanica
			Poa	cookii
			Alien Poa	annua
Marion	150	46	Indigenous	
			Agrostis	magellanica
			Poa	cookii
			Alien Agropyron	repens
			Alopecurus	australis
			Avena	sativa
			Festuca	rubra
			Holcus	lanatus
			Poa	annua, pratensis
South Georgia	3756	54	Indigenous	
			Alopecurus	magellanicus
			Deschampsia	antarctica
			Festuca	contracta
			Phleum	alpinum
			Poa	flabellata

Table 7.3. (continued)

Location	Approx. area (km^2)	Lat. °S	Genera	Species
			Alien Agropyron	repens
			Agrostis	capillaris
			Alopecurus	geniculatus
			Avena	fatua
			Deschampsia	caespitosa
			Festuca	rubra
			Lolium	perenne, temulentum
			Nardus	stricta
			Phleum	pratense
			Poa	annua, pratensis, trivialis
Crozet	505	45	Indigenous	
			Agrostis	magellanica
			Deschampsia	antarctica
			Festuca	contracta
			Poa	cookii
			Alien Holcus	lanatus
			Poa	annua, pratensis
McDonald	–	53	Indigenous Poa	cookii
Heard	–	53	Indigenous	
			Deschampsia	antarctica
			Poa	cookii, kerguelensis

Figure 7.1 shows the nesting sites of 17 species of birds (from Woods, 1970). Given its role in relation to bird populations, a conservation policy for this grass clearly has implications for bird life although an additional consideration is the presence of imported mammal predators. Figure 7.2 illustrates the growth habit of tussac grass that occurs on these sub-Antarctic islands.

The foregoing seven examples relate to the grasses' contemporary distributions. These, in turn, can be conveniently, and perhaps convincingly, explained only palaeologically.

More recently preserved grasses

Not surprisingly, more recently living grasses have been found in a wide variety of sites and preserved in different ways. The list of publications is considerable, many of which relate to the origins of agriculture.

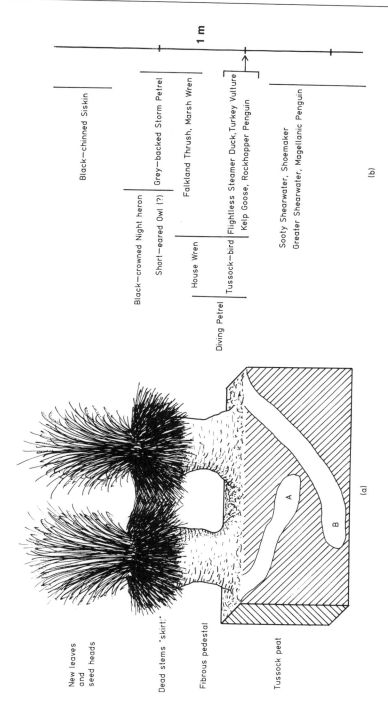

Fig. 7.1. (a) Tussac-grass profile, with diagrammatic nest burrows of sooty shearwater (A) and magellanic penguin (B). (b) Vertical zonation of nest sites in tussac grass. The sites indicated correspond to the tussac-grass profile.

Fig. 7.2. A *Poa flabellata* 'tussac grass' plant surrounded by the remnants of other tussac grass plants killed off by overexploitation.

1.   East African lake bed core samples

Reference was made earlier to drill core samples from the Niger delta (Morley and Richards, 1993). More recent material carbon-14 dated from 28,000 or less years ago has been obtained from Lake Elmenteita, Kenya (Palmer, 1976). Charred fragments of grass epidermis, the results of savanna fires, were matched to modern grasses, notably the savanna genera *Hyparrhenia* and *Themeda*, a degree of accuracy very unlikely with phytoliths and impossible with pollen. Even so, on this basis, identification to species is unrealistic. For this, whole plant detail or well-preserved flowers and fruits are normally necessary. Even then identification will be tentative since, even if on anatomical grounds close similarity is evident, it is possible that the fossil material may represent a now extinct species.

2.   Carbonised grains

If the original plant tissue has been impregnated by inorganic material or charred and then preserved in some way the term 'fossil' is appropriate. If, though, material is preserved by natural dehydration or represents plant debris that has survived cooking in proximity to a hearth, such a find is more 'prehistoric' than fossil. It has a close resemblance to living tissue but is of course inviable. Excavations of many ancient settlements for such materials have contributed to our understanding of the origins of agriculture. Of the many examples that could be chosen, one is given detailed consideration, namely maize, in Chapter 12.

Pointers to Grass Evolution

Since the grass fossil record is so fragmentary, we are left to make inferences about evolution from living materials. The 'Protopoaceae' is merely an idea and unlikely to materialise by the finding of some fossil specimen.

At various points in this book bamboos are thrown into prominence and as agrostologists explore significant parts of the world's vegetation so new and strange bamboos are discovered. Söderstrom, collecting for example in Brazil in 1976, identified a bamboo, from vegetative material, as probably *Arthrostylidium*. Shortly after, flowering specimens were found which indicated the plant to be both a new and an unusual genus, subsequently named *Alvimia* with, so far, three known species (Söderstrom and Londoño, 1988).

What makes this situation of particular interest is that it occurred in eastern Brazil in the Bahia region. *Alvimia* occurs in an area of forest that contains other bamboo genera including *Atractantha*, *Criciuma*, *Diandrolya*, *Eremitis*, *Eremocaulon*, *Piresia* and *Sucrea*, some of which are only recently recognised genera, and among which there is considerable diversity.

Earlier, Söderstrom and Calderon (1980) had raised two issues, quoted at the beginning of this chapter. These were, firstly, whether Bahia was a region harbouring relicts of ancient grasses and, secondly, whether parts of West Africa might contain related supposedly primitive bamboos since, prior to continental drift, Bahia had been contiguous with the Cameroon region. In particular, did the genus *Alvimia* with its unusual fleshy fruit have a similar counterpart in West Africa?

Although China is well known to have large numbers of bamboo species, Söderstrom's contribution has been to help shift the search for likely bamboo origins into a relatively restricted portion of the New World. It comes as no surprise of course that Söderstrom and Londoño (1988) deplore the destruction of forests in this part of Brazil on grounds both of their beauty and because of the central interest such areas have for our study of the grasses. It is indeed regrettable that the excellent example set by the island of Madeira in protecting areas of outstanding botanical interest is not more closely copied elsewhere in the tropics and subtropics.

Photosynthesis 8

A process common to green eukaryotes is the fixation of carbon from carbon dioxide in the presence of chlorophyll and involving the energy of sunlight. It is shown diagrammatically in Fig. 8.1 and among flowering plants, for example, is a regular feature of leaf mesophyll and palisade tissue. Since the first detectable metabolic product is phosphoglycerate, a compound with three carbon atoms, the process is described as C_3 photosynthesis.

Among some 15 plant families with tropical representatives there is added a second kind of photosynthesis. This is C_4 photosynthesis, so called because the first detectable product has four carbon atoms. In terms of plant organisation, C_3 photosynthesis is displaced from the mesophyll to the bundle sheath tissue surrounding the vascular bundles. C_4 activity is confined to leaf mesophyll.

Initially, there was thought to be only one C_4 process but others have been discovered as indicated in Fig. 8.2. Depending on how these alternatives are defined there are three major alternatives or up to eight or more.

Three aspects of grass biology interrelate here. These are leaf anatomy, biochemical pathways and ecological distribution. Assembling the vast body of information on grass photosynthesis, involving many investigators, has required the most detailed examination of all three aspects and some relatively consistent associations have emerged. Occasionally some inconsistency has stood out and generated a reassessment. Overall, however, different leaf anatomies, biochemical pathways and ecological distributions fall collectively into groups and prompt fundamental questions about how C_4 photosynthesis might have evolved. For example, are the different kinds of C_4 photosynthesis in some kind of sequential relationship or are they essentially independent developments? Again, how far is it possible to identify possible causes and does any projected change such as rising CO_2 level, global warming or even increased ultraviolet radiation reaching us

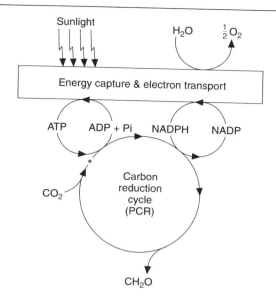

Fig. 8.1. An outline of the photosynthetic carbon reduction cycle found in C_3 plants. The point marked *
is the carboxylation reaction: ribulose bisphosphate $+ CO_2$ produces phosphoglycerate: a 3-carbon compound, hence C_3 photosynthesis. Adapted from Robinson and Walker (1981).

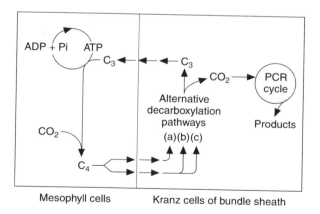

Fig. 8.2. An outline of C_4 photosynthesis. CO_2 is taken up by mesophyll cells and incorporated into 4-
carbon compounds (hence C_4). These compounds are translocated to the bundle sheath cells where they
are broken down (decarboxylated) by alternative pathways labelled a, b and c and referred to in the text.
The released CO_2 is then taken up by the conventional PCR cycle. Adapted and simplified from Edwards
and Huber (1981).

through a thinning ozone layer presage further, presumably adaptive, change? Finally, since photosynthesis is primarily an energy-capturing process, to what extent can its products serve as fuel to augment or replace our diminishing, non-renewable, fossil fuels such as coal, natural gas and oil? These are themselves products of earlier photosynthetic energy captures.

C_3 Photosynthesis

It is not the aim of this chapter to make a detailed biochemical assessment of photosynthesis generally. For this purpose see, for example, Robinson and Walker (1981), Hay and Walker (1989) and Lawler (1993), and for grasses particularly, Hattersley and Watson (1992). The aim of the present chapter is to set photosynthesis within the context of grass ecology, evolution and taxonomy. With photosynthesis more than perhaps other aspects of grass science abbreviations and technical terms have proliferated. For these the reader is referred to the Glossary.

C_3 photosynthesis includes the Benson–Calvin cycle named after its discoverers. A feature of photosynthesis is that CO_2 combines with ribulose bisphosphate. If, however, CO_2 concentration is low ribulose bisphosphate will combine with oxygen to form a compound which is eventually broken down by respiration releasing CO_2. Such 'photorespiration' constitutes 'wastage' of CO_2 and is an important point of comparison with C_4 photosynthesis discussed later.

C_4 Photosynthesis

The closer study of C_4 photosynthesis among grasses eventually revealed three contrasted metabolic pathways – see a, b and c in Figure 8.2 – each involving a different decarboxylating enzyme.

These are referred to as NADP-ME (nicotinamide adenine dinucleotide phosphate co-factor to malic enzyme), NAD-ME (nicotinamide adenine dinucleotide co-factor to malic enzyme) and PCK, sometimes referred to as PEP or PEP/CK (phosphoenolpyruvate carboxykinase).

It appeared, for a time, that each of these C_4 alternatives had a distinctive bundle sheath arrangement and so it seemed a relatively straightforward process to make a transverse section of a leaf, observe which category of bundle sheath was present and from this infer which biochemical pathway was in place.

This proved to be an oversimplification and the need for 'retyping' became recognised (Prendergast *et al.*, 1986, 1987), prime importance being given to the biochemical pathway and the bundle sheath arrangement having a secondary significance. What then emerged was that some types, predicted on the basis of leaf anatomy to be PCK (PEP), in fact were biochemically NAD-ME. Others similarly predicted to be PCK were shown to be NADP-ME.

Two important results followed from this reappraisal. The first was the recognition that NADP-ME types (biochemically) were commoner under moist conditions, and NAD-ME types in drier situations. PCK also tended to be more closely associated with drier regions. These relationships do not however hold invariably. The second result was that it became possible to

Table 8.1. Anatomy and biochemical differences among C_4 grasses. Modified from Hattersley and Watson (1992).

Key features	Sample groups
1. **'Classical'** NADP-ME; XyMS − , centrifugal chloroplasts lacking grana, cell outline uneven, suberised lamella, lacking mestome sheath	PANICOIDEAE (Andropogoneae) e.g. *Agenium, Andropogon, Sorghum, Zea* (Paniceae) e.g. *Arundinella, Panicum*
2. **'Classical'** NAD-ME; XyMS + , centripetal chloroplasts with grana, cell outline even, lacking suberised lamella, mestome sheath, with suberised lamella	CHLORIDOIDEAE *Astrebla, Buchloë, Cynodon, Leptochloa, Tetrachne* PANICOIDEAE *Panicum, Yakirra*
3. **'Classical'** PCK; XyMS + , centrifugal/evenly placed chloroplasts with grana, cell outline uneven, suberised lamella, mestome sheath, with suberised lamella	CHLORIDOIDEAE *Chloris, Dactyloctenium, Leptochloa, Spartina, Zoysia* PANICOIDEAE *Brachiaria, Panicum, Urochloa*
4. **'Arundinelleae'** as for 1. Classical NADP-ME except for 'distinctive' cells between vascular bundles	PANICOIDEAE (Paniceae) e.g. *Achlaena, Arundinella, Dissochondrus*
5. **'Neurachne'** (NADP-ME or PCK) XyMS − , centrifugal chloroplasts with grana, cell outline even, suberised lamella	PANICOIDEAE *Neurachne, Panicum* (see Fig. 8.3)
6. **'Aristida'** (NADP-ME), XyMS − , centrifugal chloroplasts without grana, cell outline even, non-suberised	ARUNDINOIDEAE *Aristida*
7. **'Triodia'** (NAD-ME), XyMS + , centrifugal chloroplasts with grana, cell outline even, mestome sheath, with suberised lamella	CHLORIDOIDEAE *Monodia, Plectrachne, Symplectrodia, Triodia*
8. **'Eriachne'** (NADP-ME), XyMS + , chloroplasts centrifugal or centripetal with grana, cell outline even (usually)	PANICOIDEAE *Eriachne*

Note: Some genera have representatives in more than one group (e.g. *Arundinella* and *Leptochloa*), conspicuously so in *Panicum*. Conversely subfamilies can share representatives of particular groups.

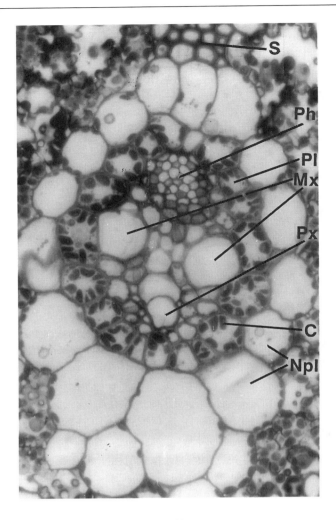

Fig. 8.3. Vascular bundle of *Panicum prionitis* × 580.
C chloroplast; **Mx** metaxylem; **Npl** non-photosynthetic layer; **Ph** phloem; **Pl** Photosynthetic layer; **S** sclerenchyma.
Note that in this case there is a double bundle sheath (**Pl** and **Npl**) and that it is the *inner* layer which is photosynthetic. Since here a non-photosynthetic layer is *not* interposed between the photosynthetic layer and the vascular bundle it is regarded as XyMS — even though the sheath is two layered. This type of bundle arrangement is classified as 5 'Neurachne' (see Table 8.1) although, in the example shown, the chloroplasts are not conspicuously centrifugal — i.e. against the cell wall away from the bundle. (Photo courtesy of F. Holt and S. Reardon).

subdivide the three major types depending partly on intricate details of leaf anatomy. Table 8.1 summarises examples of C_4 types.

Distribution of C_3 and C_4 in Poaceae Among the Subfamilies

The five major subfamilies were introduced in Chapter 2 and all occur on each of the continental land masses except Antarctica. From this might be drawn the provisional conclusion that each subfamily had differentiated before Pangaea had appreciably fragmented. To this can be added a further consideration, namely that, as explained in Chapter 7, the climatic vicissitudes thereafter could vary for each continent. Equatorial America, for example, experienced long-term stability relative to, say, Australia, while Antarctica is an extreme example of a change to almost total hostility to plant colonisation and growth.

One can envisage a situation where the grass flora as seen today of a particular continent represents an interaction between the original, founder material and episodes of climatic change – in the case of Antarctica eliminating virtually every grass except, on the mainland, one tenacious pooid, *Deschampsia antarctica*. India on its northward track became progressively more tropical, and then was eventually halted by the collision that uplifted the Himalayas. Australia seems gradually to have undergone changes from forest to desert over much of its surface coinciding with the emergence of grasses adapted to its increasingly arid conditions. While therefore a grass may be, as to its features, recognisably panicoid for example, those indigenous to this or that land mass need not and probably will not have shared the same climatic experiences through their subsequent evolution. In the case of the Bambusoideae and the Pooideae it appears that, regardless of whatever selective stresses they have been exposed to over their long history, they seem not to have the potential to generate C_4 photosynthetic variants.

Arundinoideae contains representatives of both C_3 and C_4 genera although the subfamily is referred to as 'primitive' by Renvoize and Clayton (1992). It is therefore especially interesting that three C_4 genera, *Aristida* (± 250 spp.), *Centropodia* (4 spp.) and *Stipagrostis* (± 50 spp.), occur here and that, as to C_4 mechanisms, the three genera differ from each other. Perhaps these are, therefore, each independent innovations and represent a 'new lease of life' for part, at least, of a declining subfamily.

With one exception, namely *Eragrostis walteri*, the Chloridoideae are exclusively C_4, either NAD-ME or PCK. On this basis, *E. walteri* could be regarded as a surviving 'primitive' or, perhaps more likely, a rare case of reversion from C_4. (In addition to a frequent but not invariable association with more arid habitats many chloridoid grasses are also salt tolerant.)

Table 8.2. Watson (1990b) recognises in the subfamily Panicoideae, two supertribes (cf. Clayton and Renvoize, 1992).
Using the former taxonomic arrangement, the distribution of
photosynthetic types is shown.

Supertribes	
Panicodae	Andropogonodae
PANICEAE	MAYDEAE
C_4 (NAD-ME, PCK, NADP-ME) and C_3	C_4 (NADP-ME)
ARUNDINELLEAE	ANDROPOGONEAE
C_4 (NADP-ME)	C_4 (NADP-ME)
NEURACHNEAE	
C_4(NADP-ME),	
C_3, C_3–C_4	
ISACHNEAE	
C_3	

The situation with the Panicoideae is more complex. Using the
taxonomic arrangement for this subfamily of Watson (1990) it can be set
out as in Table 8.2.

Two tribes Neurachneae and Paniceae deserve comment here. The first
contains three genera *Thyridolepis* (C_3), *Paraneurachne* (C_4) and *Neurachne*
(C_3, C_4, and C_3/C_4), this last being 'intermediate' and considered later in
some detail. The tribe Paniceae has, as its centrepiece, the genus *Panicum*
with about 400 species. Hattersley (1987) lists for this genus NAD-ME 14
spp., PCK 1 sp., NADP-ME 7 spp., and in addition there are numerous C_3
types and many others yet to be classified. *P. hians* = *P. milioides* is another
C_3/C_4 intermediate.

Other Instances of Photosynthetic Pathways in Relation to Taxonomy

1. *Alloteropsis semi-alata.* This species, introduced in the previous chapter,
is remarkable in that it contains C_3, and three versions of C_4. Not only is it
the most internally diverse species in terms of photosynthetic mechanisms
but it is present on drifting continents that have differed so much in their
climatic histories.

In East Africa, Gibbs Russell (1983) found populations different for photosynthesis within 100 m of each other, although no evidence of hybridisation was reported. Earlier, however, Ellis (1981) had found examples in

Tanzania, Zambia and Zimbabwe with intermediate $\delta^{13}C$ values (-24.24 to $-20.70‰$). Although Hattersley and Watson (1992) identify the situation as 'the most interesting yet documented for C_3/C_4 intermediates among angiosperms' and raise the prospect of contemporary evolution of photosynthetic pathways in these three countries, the matter remains tantalisingly unexplored.

2. *Bouteloua curtipendula*. This grass, found naturally occurring in subtropical and warm-temperate parts of the New World appears to be an NAD-ME/PCK intermediate as judged by metabolic activity.

3. × *Cynochloris* is an Australian naturally occurring hybrid between two chloridoid genera, *Cynodon* (NAD-ME) and *Chloris* (PCK). The hybrid shows both types of enzyme activity, Prendergast *et al.* (1988).

4. *Oryzidium*. This example is interesting because although it is NAD-ME, a pathway associated with drier habitats, its single species floats in water. It occurs in Zambia.

5. *Yakirra* is a genus recognisably close to *Panicum* and only recently distinguished from it. Its NAD-ME photosynthesis might well be seen as derived in a *Panicum* exposed to increasing aridity since it is confined to Australia.

These five examples indicate that a given single species can contain all four modes of photosynthesis (*Alloteropsis semi-alata*), that two types of activity can co-exist in a single plant (*Bouteloua curtipendula*), occasionally different C_4 mechanisms will occur in hybrid combination (× *Cynochloris*) and certain genera prove exceptional to such associations of NAD-ME with drier habitats (*Oryzidium*). It must also be recognised that *Cynodon*, a chloridoid for example, although it shares the NAD-ME pathway with *Yakirra*, a panicoid, is not closely related. Similarly *Panicum* species that differ by being C_3 or C_4 NADP-ME are in a taxonomic sense more closely related than *Panicum* and the andropogonoid grasses that are, uniformly, NADP-ME.

Modifying Taxonomy?

A problem is to decide what 'weight' to give to contrasts in photosynthetic pathways. Suppose, for example, *Panicum* is in this sense a 'versatile' genus. This could be 'biologically' the correct interpretation. A quite different impression is created if all C_3 *Panicum* species are separated to *Dichanthelium* and all PCK species to, say, *Brachiaria*. Again if *Alloteropsis* were split into different species for the same reason our perception of the *Alloteropsis* 'complex' is subtly altered. Since the resources available are insufficient to explore all the interesting situations we must make do in most cases with provisional taxonomic judgements. The underlying issue is how such situations evolved and it is at this point the 'intermediate' C_3–C_4 types are seen to be of possible significance.

C_3–C_4 Intermediates

The following discussion begins with the primacy of the biochemical pathway and regards anatomical considerations as not unimportant but clearly secondary whether supportive or otherwise. Recalling the parameter $\delta^{13}C$ introduced in Chapter 7, C_3 plants discriminate against $^{13}CO_2$ relative to $^{12}CO_2$ but C_4 plants do not. C_3 plants tend to have $\delta^{13}C$ values of -22 to $-35‰$ while C_4 plant values are in the region -9 to $-18‰$. Especially low values ($-18‰$ and beyond) in nominally C_4 plants or especially high values down to $-22‰$ in nominally C_3 plants could indicate possible intermediacy. Additionally, it will be recalled, low CO_2 concentration in the surrounding atmosphere would favour C_4 photosynthesis.

Panicum milioides shows, at lower CO_2 values, an ability to discriminate greater than at higher values. Most of the CO_2 is fixed in the mesophyll by Rubisco. A smaller part is fixed in the bundle sheath with some CO_2 derived from glycine decarboxylation. Since the bundle sheath is not gastight, Rubisco here *again* discriminates against $^{13}CO_2$. At low CO_2 partial pressures there is thus *more* discrimination by C_3–C_4 plants using a 'glycolate shuttle' than in normal C_3 plants. Under such circumstances the $\delta^{13}C$ value is *not* helpful in identifying C_3–C_4 activity.

The parameter Γ (gamma) is a measure of the 'CO_2 compensation point', the balance between photosynthesis and total respiration. The synthesis of C_4 compounds uses phosphoenol pyruvate carboxylase as the initial step and incorporates CO_2 without competition from oxygen. By contrast, Rubisco (ribulose bisphosphate carboxylase) will combine with O_2 if CO_2 partial pressures are low. If, therefore, 4-carbon compounds are broken down in the bundle sheath, thereby creating a relatively high CO_2 concentration, the Rubisco works 'efficiently'. By comparison, C_3 plants operating Rubisco in the mesophyll and thus lacking a 'concentrating mechanism' are more prone to its combining with O_2 rather than CO_2 eventually leading to CO_2 output via photorespiration. Under bright light therefore, C_3 plants have Γ values of 2–4 mmol m^{-3} while for C_4 plants Γ is from 0.1–0.4 mmol m^{-3}. If an intermediate value were found it could imply a less than fully functional C_4 mechanism.

Neurachne minor is an Australian grass and one shown to have Γ values from 1.2 to 1.8 in plants transferred from the field and 0.3 to 0.5 in new leaves of cabinet-grown plants. *N. minor* is therefore interesting because it combines the low $\delta^{13}C$ values of a thorough-going C_3 plant with the low Γ values of one that is C_4. It is noteworthy that Γ values are lower for *N. minor* than *P. milioides* implying that, in this respect, *N. minor* is more C_4-like (Hattersley and Stone, 1986).

A drawback to work with both of these species is that neither will cross with its near relatives having different photosynthetic pathways (C_3 *N. alopecuroidea, lanigera, queenslandica* and *tenuifolia*, C_4 *N. munroi* quoted by Hattersley and Watson, 1992) and therefore their physiological differences cannot be subjected to genetic analysis. It is this consideration which underlines the potential interest of *Alloteropsis semi-alata* where it seems reasonable to expect the possibility of making within-species crosses artificially among representatives of different photosynthetic pathways.

Evolutionary considerations

The evidence available points to C_3 as the earlier and C_4 as the later derived types of photosynthesis. If this is so, how did C_4 photosynthesis arise? It is difficult to imagine an 'instant' changeover since a group of features would have to be in place. The interest of *Panicum milioides* and *Neurachne minor* is that they establish evidence of intermediacy and thus the possibility of a gradual transition.

In whichever way C_4 photosynthesis evolved and although the order of events is unclear the following changes were among those required.

1. Activation of the appropriate decarboxylating enzyme.
2. An anatomical alteration to vascular bundles invested with chlorenchyma.
3. A shift of C_3 activity to the bundle sheath.
4. A switch in mesophyll activity.

There does not seem to be an infinite number of C_4 alternatives but only perhaps eight or so that, significantly, can be shared out among either related or unrelated plants. The likelihood of independent origins in related groups in different parts of the world seems overwhelming. Where among closely related grasses such as the andropogonoids sharing NADP-ME one pathway is common to all of them, the simplest explanation is that they diverged from a common source *after* it had adopted that form of photosynthesis.

What then was the trigger for the evolution of C_4 photosynthesis? Did a conspicuous though localised aridity create the conditions for the emergence of NAD-ME types? Was this in or out of step with a rise in temperature elsewhere that caused the emergence of NADP-ME types? Or, was there a decline in global CO_2 that in areas sufficiently warm prompted increase in the efficiency of its use?

One might usefully compare here the genus *Panicum* in Panicoideae with the entire subfamily Chloridoideae. The co-existence today in *Panicum* of C_3, NAD-ME, PCK and NADP-ME argues that it was long widely dispersed and in place ecologically to react appropriately to the need for C_4 innovations. By contrast, the relative uniformity of the Chloridoideae and a commitment to NAD-ME/PCK suggests rapid evolution *after* the arrival of appropriately arid conditions.

> The uniformity which runs through the subfamily Chloridoideae suggests strong evolutionary coherence of the genera and probably reflects the 'suitability' of the C_4 pathway to the environment to which the subfamily is adapted.
>
> Renvoize and Clayton (1992)

No obvious candidate as progenitor is recognisable for the Chloridoideae and one possibility is that if C_3 'proto chloridoids' were dispersed in humid regions that subsequently became arid they were rapidly outcompeted by their C_4 successors and became extinct.

Another possibly instructive comparison is between (say) the supertribe Andropogonodae and the genus *Aristida* occurring in a different subfamily Arundinoideae. The NADP-ME pathway is the common feature but the leaf sections differ in having in *Aristida* a *double* chlorophyllous bundle sheath. Whatever common ancestor the Andropogonodae might share (suggesting a single NADP-ME innovation), the *Aristida* situation surely indicates the possibility of independent origin.

As to when these events might have occurred, the known or likely influences were indicated in the previous chapter. That is to say C_4 evolution was perhaps possible during the Palaeocene with diminished CO_2 values but more likely in the Miocene with its, additionally, hotter more arid climates.

Some Contemporary Implications of Photosynthesis

Walker (1992) showed that of the energy reaching the earth's surface at the latitude of the United Kingdom the maximum that could be fixed by photosynthesis would be about 4.5%. In practice, due to a combination of factors such as pests, diseases, weeds and periods of drought, the value would be about 1.0%.

Before assuming, therefore, that farming is inefficient, it is worth pointing out that Canada, for example, produces just under 60 million tonnes of grain together with about 200 million tonnes of straw, chaff, stover and cobs. Assuming one third of the 'non-grain' material were used on farm the remaining two thirds if converted to ethanol would provide, at present rates, *all* of Canada's fuel requirements (Wayman and Parekh, 1990). We shall however return, subsequently, to this example.

Further south, in the tropics, *Saccharum officinarum*, sugar cane, is harvested as thick culms and pressed through heavy rollers to extract the juice. The three important products are cane sugar, molasses (the brown sugary residue) and bagasse (the fibrous remnants). Since all three are the results of photosynthesis they represent energy capture and potentially, therefore, could be used for fuel, the sugar and molasses being fermented to ethanol and the bagasse burned.

A similar rationale applies to *Sorghum bicolor*, sorghum in its 'milo' or sweet form, except that, additionally, it yields a grain crop and is adaptable further north, a situation increasingly exploited in the United States.

Although, for a century or more, bagasse has been used on site as a fuel for sugar factories, the amount produced is vast and is a familiar waste product seen heaped up and traditionally either neglected or regarded as a nuisance. The situation began to change when in the 1970s the Organisation of Oil Producing and Exporting Countries (OPEC) dramatically raised the price of oil. Given that oil is a non-renewable fossil fuel, the impetus to develop renewable fuel sources was apparent. By 1977 Gopalakrishan and Nahan had argued, for three Hawaiian islands, circumstances were such that, potentially, bagasse from their sugar factories could provide 75% of the public power supply requirements and contracts between sugar producers and the supply company were in some cases already operating. By 1990 in Brazil some 2.5 million automobiles were fuelled by alcohol derived from sugar and the task of replacing all imported oil was in sight (Wayman and Parekh, 1990). Fuel from 'biomass' was not so much a theory as a practice driven by economic reality.

More recently, Therdyothin *et al.* (1992), examining electricity generation from Thailand sugar mills, offered the following analysis. Four options were identified: (a) burning bagasse, (b) burning bagasse plus cane trash from the field, (c) use of a single extraction–condensing steam turbine and (d) use of a double condensing system, these representing increasing degrees of sophistication. The background to the study is instructive. Given the increasing electricity demand in Thailand, about 12% per year, national investment in new power plant and the purchase of fossil fuel threatens to become an excessive burden. The policy therefore of 'cogeneration', that is power generation by private companies and sale of surplus to the national grid, is an attractive option. This in turn creates a shift in attitudes. There is an incentive for sugar manufacturers to improve the efficiency of a factory so as to generate saleable electricity and an incentive for the government to set an imaginative pricing framework within which it will buy the power. Since the sugar milling season does not extend throughout the whole year, some ingenuity is required to extend the electricity producing aspect of a sugar mill so as to have, profitably, electricity to sell out of season.

Clearly, a sugar manufacturer setting up a new factory or re-equipping an old one could come to see electricity generation as an increasingly important part of his activity and a major consideration directing investment and management policy.

Biomass versus 'set-aside'

In Western Europe a more or less stable population combined with rising crop productivity has led to the accumulation of surplus. While such 'grain mountains' provide a famine reserve for emergencies elsewhere, to give away, on a regular basis, food to poor countries both undermines farming there and creates an unwelcome and vulnerable dependency in the recipients. The European response, and similarly in North America for the same reasons, has been to withdraw land from agriculture in the policy of 'set-aside' and to compensate the farmers in some measure for crops foregone.

Given certain safeguards, conservationists have generally welcomed set-aside since a brake on intensive farming has meant regeneration of scrubland habitats, a diversification of the flora and, as a consequence, encouragement of insects, birds and other animals. There are, however, other considerations. If one assumes further enlargement of the European Union and with it an extension of present trends for population and production, there could eventually be between 50 and 80 million hectares withdrawn from cultivation (Grassi and Bridgewater, 1992). Even the most enthusiastic conservationist might begin to have doubts about all of this land reverting to wild nature especially if some of it could be developed for biomass and thus fuel production to reduce our dependence on fossil fuels.

How far, therefore, is it feasible to replicate the successes of the sugar industry in the much less favourable climates of Europe?

The Ideal Biomass Plant?

The basis of a renewable resource for fuel using biomass is the need for a highly efficient crop that, in a short time, can yield harvestable material to be either burned or fermented under controlled conditions. Among candidate species several grasses, short-generation coppiced willow and oil seed rape have been tested, and work continues on all of them.

Speller (1993) defined the properties required in an ideal biomass species. There should be:

1. Dry harvested material for efficient combustion.
2. Perennial growth to minimise establishment and growing costs and to lengthen the growing season to exploit photosynthetic activity for as long as possible.
3. Good pest and disease resistance.

4.   Frost hardiness.
5.   Efficient conversion of solar radiation to biomass energy. (C_4 species have a theoretical maximum of about 55 t ha^{-1} compared with 33 for C_3 species).

On this basis grasses belonging to the genus *Miscanthus* could seem closely to approach the ideal.

Miscanthus

This genus of about 20 species is centred on South East Asia but some representatives are hardy in northern Europe. *M. sinensis* occurs in various forms and 'Zebrinus' is a familiar ornamental.

Miscanthus, a C_4 grass, belongs to a group of andropgonoid grasses that includes *Eranthus* (28), *Imperata* (8), *Miscanthidium* (6–7), *Miscanthus* (20), *Narenga* (2), *Sclerostachya* (2–3) and *Sorghum* (30), all related to *Saccharum* and able to form hybrids with it.

Two species, *M. sacchariflorus* and *M. sinensis*, have attracted considerable research interest in Europe, and in Germany for example some 200 ha are under pilot cultivation (Speller, 1993). Yields vary from 11.7 to 25.3 t ha^{-1} yr^{-1}. Taking a median value of 18.5 t ha^{-1} yr^{-1} this represents, at 16% moisture capacity, 277,500 megajoules, or at 100% conversion efficiency, 77,083 kilowatts. In practice conversion efficiency might (optimistically) be nearer 50% or about 38,500 kilowatts. What, in theory, might this mean applied to existing commercial activity?

To return for a moment to the Canadian example, those 60 million tonnes of grain and that associated straw are presently sown, raised, protected against pest and disease, fertilised and harvested through the use of fossil fuel. What point is there in assembling such biomass if non-renewable fuel sources were needed to create it? While one cannot establish a crop on the basis of zero energy, the energy output must substantially exceed the energy input from fossil fuel. For *Miscanthus sacchariflorus* the energy value after harvest is about 30:1 relative to establishment (J. Kirkpatrick, pers. comm.) although this does not take account of energy required to deliver it to a processing plant.

An example from aviation

To convey a 747-400 jumbo jet with 386 people on board together with freight, luggage and mail from London to Hong Kong requires about 166 tonnes of fuel or 5,300,000 megajoules (Somerville, pers. comm.). On the assumption that a hectare of *Miscanthus* would yield 277,500 MJ some

19 ha would be required. Since efficiency conversion would be excellent at 50% and neglecting conveyance of fuel to the aeroplane, 38 ha would be a more realistic figure. Due to the jetstream, the return flight would be more energy demanding needing 187 t of fuel or 5,990,000 MJ. At an optimistic 50% conversion efficiency this represents 44 ha of *Miscanthus* or more than 80 for the round trip.

Further reflections

The aviation example can be interpreted in two ways. One could hastily dismiss fuel from biomass as hopelessly inadequate to meet our needs. Alternatively it can underline our immense and profligate requirement for fossil fuel which sooner or later will become too expensive to extract and renewable resources will have to command more interest. Even so, for Europe at present levels of fuel requirement, if 25% of arable land were given over to biomass production it would only yield about 5% of energy requirement (McMullan *et al.*, 1983).

Aircraft engines are becoming more efficient. If it were indeed feasible to convert biomass to aviation fuel, a situation not yet realised commercially, it might be more economical to refuel aeroplanes at their most equatorial landing sites when fuel from sugar becomes a more feasible proposition.

The most encouraging feature is that through photosynthesis energy sources *are* renewable. What is obvious is that energy policy needs to be fundamentally rethought. Ultimately we have no option.

Contrasted Life-Styles   9

There is a descending hierarchy from panicle through spikelet, floret and embryo sac which at every level can manifest adaptive alternatives. Again, most grass plants rather than being isolated single specimens are members of a population that can be either inbred, outbred or show subpopulations that differ in this respect. Populations can demonstrate long-term occupancy of some habitat through perennialism or repeated annual generation or their presence can be relatively transitory in some ecological succession. Many managed habitats maintain the predominance of grasses. For example sports turf, amenity areas, motorway embankments, permanent pasture and fields of annual cereals are all examples of deflected or arrested climaxes. Again, the same species, perhaps as different ecotypes, can occur in more than one habitat and react differently, for example, to levels of grazing. Finally, of course, few grass populations consist of one species only but co-exist both with other grasses and a range of other, mostly angiosperm, species that can provide competition, protection and alternative hosts for shared pathogens and pests.

A Possible Prototype: The Seedy Perennial

Among angiosperms, woody perennial forms have long been considered more primitive with the adoption of herbaceous perennial and ultimately annual and ephemeral deriaties as end points. While this might make bamboos seem an obvious starting point, received opinion is that their woody stems are 'specialised' and therefore secondarily derived. Figure 9.1 provides a basis for discussion. The asterisk indicates a theoretical starting point though a movable one depending on the inclination of the reader.

Ability to produce seed can lead to acquisition of a habitat and perennial growth to retention and extension of hold upon such a habitat. A

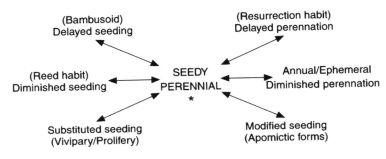

Fig. 9.1. Modifications of the seedy perennial.

habitat changing rapidly over time, such as an arable field, puts seediness at some advantage but the matter is not clear cut since rhizomatous weeds such as *Agropyron repens* can flourish. The reed habit in *Phragmites australis* is a curious situation since, despite abundant flower production, seed formation can be rare with dependence on clonal propagation of populations over many hundreds or even thousands of years implying the existence of a well and widely adapted genotype resistant to competition, whether from within or beyond its own species. An alternative to this exists and both are considered in the later discussion of *Phragmites*. Overall, the (slightly) seedy perennial grass with its tenacious ecological hold seems an insurance against eventual ecological replacement and the other variants perhaps best regarded as derivatives of it in response to particular circumstances.

Mating Systems

Among those grasses where seed formation is conspicuously part of the reproductive strategy, the emphasis can be upon either in- or outbreeding, each secured by some modification of the flower. Figure 9.2, taking an hermaphrodite flower, able to self and cross-fertilise, as its starting point, sets out various possible modifications.

About such mating systems it is necessary to add qualifying comment since their expression against the background of a particular species can vary in subtle ways that can have profound effects.

1.   One anti-selfing mechanism can be superimposed upon or co-exist with another. In *Tripsacum*, for example, inflorescences are monoecious with proximal female and distal male spikelets. Plants are also dichogamous with conspicuous protogyny.

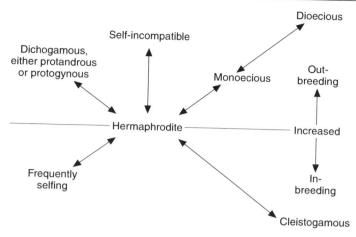

Fig. 9.2. Modification of mating system.

2. Commitment to a given mating system is neither absolute nor irreversible. *Buchloë* is commonly regarded as a dioecious grass but a given population will show occasional monoecious forms. The existence of a well defined S–Z incompatibility system in those genera where it is known to occur still allows the possibility of occasional self-fertilisation.

3. There can be a seasonal effect. *Rottboellia* is one example among several known to switch from cleistogamy to chasmogamy with increasing daylength.

4. 'Exceptional' progeny can have populational effects. *Hordeum* can be cleistogamous and in any case shows a high degree of inbreeding. Varietal mixtures grown together over many years show that chance outcrosses can generate new genotypes which can become conspicuous and thus change the character of a population.

5. Genera, especially large ones, need not show overall commitment to a particular mating system. *Poa*, for example, has among its different species those that are inbreeding, cleistogamous, with unisexual florets or apomicts. No grass species utilises all the options that exist across the family and there follow some contrasted examples.

6. The sexual system can be abandoned, partially so in facultative apomixis and completely in obligate apomixis. An extensive literature exists notably for *Cenchrus*, *Pennisetum* and *Poa*.

With these various considerations in mind the contrasted life-styles of a selection of grasses will now be considered.

Bambusoideae

Dendrocalamus strictus

The genus *Dendrocalamus*, adventive through India to the Philippines, is related to *Bambusa* and *Gigantochloa*. *D. giganteus* is probably the world's largest grass rising to 30–35 m tall. *D. strictus* is solid-stemmed and an important timber. The plant grows vegetatively for two, three or even four decades and then flowers and seeds spectacularly (Janzen, 1976). The inflorescence is interauctant with pseudospikelets. Florets have no lodicules, six anthers and a tripartite stigma. The remarkable *in vitro* flowering tech-nique developed by Nadgauda *et al.* (1990) condenses the bamboo flow-ering cycle to a few months. This would now permit a timely exploration of incompatibility in a bamboo and could resolve whether the S–Z system extends to this subfamily.

Flowering records for *D. strictus* are substantial and Janzen (*ibid.*) has shown how conspicuously variable is the inter-mast period for different specimens of the species – a fact he attributes both to human selection and ecogeographical contrasts. The extremes range from about 8 to 44 years although some doubt attaches to the lower figure.

Janzen reports an estimate of a 40 square yard clump of *D. strictus* yielding 320 lb of 'seed' or about 4.5 million caryopses. Given approxi-mately stable populations it could be regarded as wastage on an heroic scale. Over (say) a 40 year period, 39 years would produce no offspring. In the mast year all but one or two caryopses would be consumed by animals or people or simply deteriorate.

For *D. strictus* holocarpy is not inevitable and Mathauda (1952) reported that 8–27% of clumps could survive flowering.

Reproductively, *D. strictus* depends heavily on human beings fragment-ing and distributing clones. Seed dependence is made to seem almost irrelevant and yet given the evident diversity and the new techniques available there is clearly an opportunity for genetic exploitation. It may be, too, that on a lesser scale similar, hitherto unsuspected, opportunities exist in numerous other bamboos.

Arundinoideae

Micraira *species*

This account is based on Lazarides (1992). The genus *Micraira*, consisting of a dozen or so known species and some still to be described, is endemic to Australia and occurs in regions that experience seasonal aridity. Growth habit is moss-like and among grasses almost unique in having a spiral rather than a distichous phyllotaxy. Spikelets are two-flowered, lack lodicules, have two rather than three anthers, are protogynous and reproduce sexually rather than by apomixis. Six species (*M. adamsii, multinerva, spinifera, subulifolia, tenuis* and *viscidula*) have 'resurrection' properties, namely the ability to revive after dehydration to air dryness. The genus is C_3 in an area where the grass flora is predominantly C_4.

Micraira species have a disjunct distribution and are missing from habitats where they might be expected. Vegetative reproduction yields a long-lived stoloniferous mat formation with seedling establishment seemingly a relatively rare event following wind dispersal of the small unspecialised caryopses. One impression of *Micraira* is of a xerophyte with a tenacious hold upon its habitats which, since it is disjunct, can also be read as the relict of a formerly wider distribution. A problem yet to be resolved is the long-term aridity of Australia. Does it, with minor perturbations, reach back only to the Pleistocene, or to the Miocene, Oligocene or even Eocene? And of course where aridity has been long-term how much shifting and contraction has occurred? As a C_3 genus perhaps *Micraira* is relatively old but the C_4 genera *Eragrostiella, Sporobolus* and *Tripogon* have also evolved resurrection forms. Since resurrection species are not as numerous as for the Namib Desert of South Africa, Lazarides suggested there is a longer period of more extreme aridity than in Australia. (One curious aside to this is the apparent absence of resurrection grasses from the New World mentioned earlier.)

Phragmites australis – *common reed*

This genus is widely distributed throughout the temperate and tropical regions and is a familiar denizen of river and lakesides. Although it flowers abundantly, seed production can be minimal and seed establishment extremely rare. One pattern, consequently, is of a few clones each covering large areas and by implication, therefore, of great age. The paradox is that genetic recombination foregone, having been consigned largely to the remote past, seems no bar to widespread current success. Such a view of this species is reinforced by the work of Hauber *et al.* (1991) in the Mississippi delta. Here a land mass about 500 years old has been colonised by a 'background' type of *P. australis* and, interspersed among it, a 'patchy' type, and each appears to be remarkably persistent. Seed formation is extremely rare as are 'new' phenotypes.

An important study is by Björk (1967) who included information not only from Sweden but from Europe, Argentina, North America, North Africa, the Black Sea coast, India and Iraq. Using $x = 12$, he records claims for $3x$, $4x$, $6x$, $7x$ and $8x$ although his own work includes only tetraploids and hexaploids; the former he considered to be the most common worldwide ploidy level. Fertile seed production among hexaploids he found to be zero and it was highly variable among tetraploids ranging from 3 to 4325 per inflorescence.

In the Netherlands van der Toorn (1972) found seed number per panicle to be between 2200 and 5800 and commercial seedling production is significant. This author comments that seed output is lower in natural reed beds in the Netherlands but the seed number per panicle (between 30 and 1690) is far higher than most values recorded elsewhere.

In no modern conservationist sense is *Phragmites* a threatened genus although its subfamily and more particularly its tribe, the Arundineae, have sometimes been given virtually 'relict' status. One assessment might be therefore that *Phragmites* is a highly durable relict with good prospects and no obviously foreseeable demise. Perhaps three species are recognisable but one complex of surviving clones might be at least as realistic an interpretation.

Pooideae

Festuca ovina – *sheep's fescue*

Hackel (1892) considered *F. ovina* to be a single species, divided into nine subspecies and a complex array of lesser taxa. During the following century the *F. ovina* 'complex' challenged many taxonomists and generated various cross-currents of opinion. The problem primarily is that numerous variants, while clearly referable to '*F. ovina*', are none the less distinctive in morphology, appear to have some edaphic and/or geographical confinement and on closer investigation reveal cytological differences. A notable attempt to resolve the situation is that due to Wilkinson and Stace (1991) for representatives in the British Isles. It can be summarised as follows. *F. ovina* in the sense of Wilkinson and Stace is seen to consist of three subspecies, one of which is further divided. There are then several taxa not brought within *F. ovina* but regarded as separate though obviously similar species.

Figure 9.3 sets out the major features. Within this perception of *F. ovina* the two subspecies *hirtula* and *ovina* do, from place to place, co-exist in some instances. Apparently, however, each taxon retains its distinctiveness and instances of hybridisation in nature seem rare. If one regards, for a moment, the whole complex in Fig. 9.3 as a '*Festuca ovina* aggregate' then the impression is of subtle diversification preserving an approximately similar external appearance but, beneath this, the adaptation of chromosome races involving particular gene complexes to particular ecogeographical areas. Such 'cryptic' speciation might well occur in quite unrelated

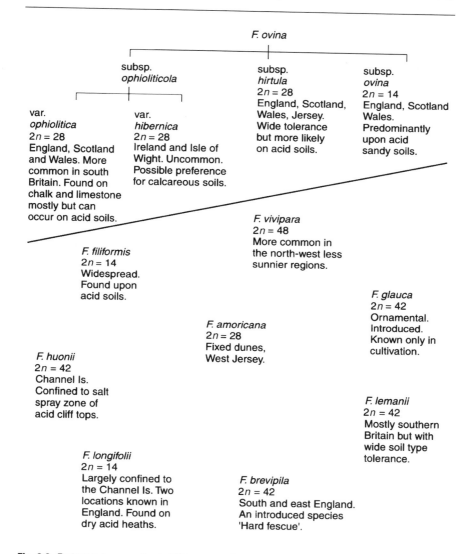

Fig. 9.3. *Festuca ovina* according to Wilkinson and Stace (1991).

grass genera elsewhere in the world and it is worth pointing out that our understanding of *F. ovina* has depended on the continuous availability of sufficient interested botanists, herbarium collections and well-recorded field distributions for more than a century.

F. ovina, in all its forms, is perennial, as are its related species shown here. Its long-term occupancy of habitats forms part of the classic studies on *Festuca* by Harberd (1962). Given that *F. ovina* is self-incompatible, if two seemingly discrete and separate plants growing in the same location yield no more seed on crossing than selfing there is a supposition that they are ramets of the same clone no longer physically connected because of the effects of vegetative reproduction through time. A quadrat of 10×10 yards was marked out and sampled in a 1600 point grid, the points being at 9 inch intervals. From each point a small circular turf was collected and if present a tiller of *F. ovina* isolated and grown on. The procedure yielded 1050 plants. From these three distinctive groups of plants, each approximately homogenous within themselves, were selected. Closer examination exposed various subtle differences leading to the conclusion in each of the three groups that more than one genotype was involved. Returning to the original sward and the competitive conditions within it showed that a few genotypes tended to be widespread and, therefore, of great age, perhaps say 1000 years. The matter is not straightforward since for each clone the style of reproductive growth was not identical and, over so long a time scale, it does not follow that ecologically each clone would have had the same relative competitive advantage. What is, however, clearly suggested is the long-term durability of particular clones rather than a pattern of short-term perennation and high diversity. An absorbing question is how far the situation Harberd describes is a microcosm of patterns that have led, over a far longer time scale, to the wider speciation described by Wilkinson and Stace.

Poa annua – *annual meadow grass*

Poa annua $(2n = 28)$, although small and seemingly insignificant, is a highly successful grass reproducing sexually and found throughout the temperate regions and in the cooler upland tropics. Although hardly an aggressive weed it is a familiar constituent of gardens and all but the most meticulously maintained temperate lawns. Part of its interest is that it can occur in either swards or as isolated specimens and with a biology modified to each situation.

MacArthur and Wilson (1967) recognised that under 'density independent' regulation, that is isolated plants not competing with surrounding vegetation, genotypes with the greatest intrinsic rates of natural increase would be selected, ('r' selection). Under 'density dependent' regulation, that is plants competing with surrounding vegetation, those with the greatest capacity for vegetative growth would preferentially evolve ('K' selection). If genetic variability for life histories occurs in natural populations then choice of material under appropriate conditions might be able to demonstrate individual plants having either r or K strategies. Law *et al.* (1977) utilised *Poa annua* to explore this topic.

Selections were made in north-west England and north Wales choosing material to represent extremes of density independent and dependent regulation which they referred to as 'opportunists' and 'pasture' types respectively. Tillers of each were grown under field conditions and their seeds sampled to provide test populations which were eventually set out in a field with half metre (i.e. non-competitive) spacing. The species is nominally annual but some plants persisted beyond the 17 month duration of the experiment. Plants of opportunist origin produced seed during their first spring 10 to 20 weeks after germination. Those of pasture origin had a more extended seed production but tended to concentrate this around the second spring. Conversely, vegetative spread was greater in the latter. Again, opportunists plants tended toward shorter life length. *Poa annua* is mostly inbreeding and this species therefore, because of this together with its prolificacy and relatively short generation time, provides a convenient model for population studies. One issue, for example, concerns 'risk'. Suppose a pasture-type seedling were established. If seed production early were too high, mortality might ensue, but if seed production were indefinitely delayed survival would depend inordinately upon vegetative spread. The compromise inherent to a particular seedling may or may not result in survival depending on how soon and how severe was the competing growth of its neighbours. Again, since inbreeding is not absolute, gene exchange between the two types is likely, this tending to dilute their differences in opposition to selection in different habitats tending to maintain them.

The frequent flowering habit in *Poa annua*, given its relatively unspecialised floral structure and the alternative r and K strategies evident makes it a convenient class subject to demonstrate some essentials of grass reproductive biology.

Alopecurus myosuroides – *black grass*

This genus contains about three dozen species including both annuals and perennials. *Alopecurus pratensis* 'Aureomarginatus' or 'Golden foxtail' is a well-known ornamental $(2n = 28)$ and propagated vegetatively. Of more particular interest is *A. myosuroides*, a significant weed of temperate agriculture. It is introduced here against its generic background.

An early study was that of Strelkova (1938) who records the genus from above the Arctic Circle in tundra near the limit of vegetation, in mountains up to 4500 m above sea level, and in Egypt, Persia, Afghanistan and the Himalayas. Nowadays it is recognised to be distributed more widely in regard to its most conspicuous species. Strelkova recognised the pattern set out in Table 9.1.

Table 9.1. *Alopecurus* after Strelkova (1938).

Species grouping	Ploidy	Geographic region	Habit
Annuae 7 spp. including A. *myosuroides*)	$2x = 2n = 14$ except A. *geniculatus* $2n = 28$	Europe, N America, Africa, Asia	Annual
Pratenses (4 spp.)	$4x = 2n = 28$	Europe, W Asia	Perennial
Ventricosae (5 spp.)	$4x = 2n = 28$ (where known)	Europe, part of Asia	Perennial
Vaginatae (8 spp.)	$8x = 2n = 56$ (where known)	Caucasus, Crimea, Turkestan	Perennial
Alpinae (7 spp.)	$14x = 2n = 98$ (2 cases) $10x = 2n = 70$ (1 case)	Arctic and montane regions	Perennial

Since this pioneer study, the genus has become better known but perception of the association of annual habit, low chromosome number and wider distribution versus perennial habit, high chromosome number and narrower ecological range remains accepted. Among the annuals *A. myosuroides* (Fig. 9.4) is, according to one's viewpoint, a pernicious weed or an outstanding ecological success. It has been studied in the British flora by Naylor (1972) where it is described as locally common and a serious weed in winter-sown cereals. Put differently, arable agriculture creates an environment more suited to *A. myosuroides* than wheat since the former is self-sustaining while the latter, through human agency, requires an annual injection of seed. The south-east of England with lower rainfall and higher temperatures is particularly prone to high-density infestations. Vegetation on land converted to 'set-aside' (that is allowed to return to scrub and eventually woodland) progressively outcompete this species although in the event of soil disturbance its seeds would, for about four years, readily germinate.

Black grass has been subject to herbicide trials and proved relatively resistant. It is most susceptible around the three-leaf stage but spraying creates problems for the accompanying cereal with which it is, developmentally, about in step. Given that a plant can produce up to, say, 8000 seeds, even reduction to one or two plants per square metre creates no real prospect of elimination although the plant has no capacity for vegetative spread. Left unchecked, a build-up to 560 plants per square metre has been

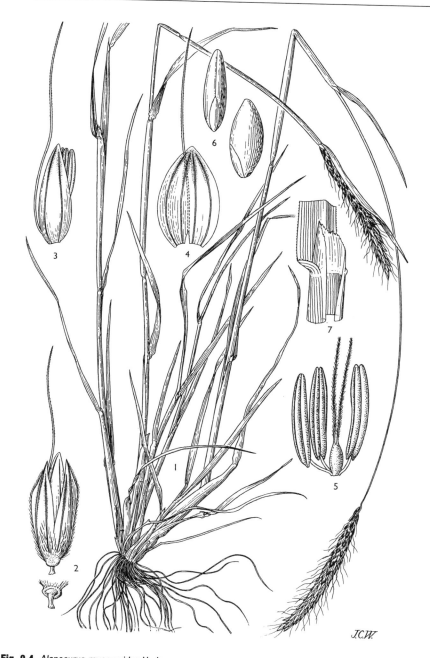

Fig. 9.4. *Alopecurus myosuroides* Huds.
1. habit $\times \frac{2}{3}$; **2.** spikelet showing glumes connate in lower third $\times$ 6; **3.** lemma and awn, side view $\times$ 6; **4.** lemma $\times$ 6; **5.** flower $\times$ 8; **6.** grain $\times$ 8; **7.** ligule $\times$ 6.

recorded. *A. myosuroides* provides an example of a widespread plant, not especially common but, given arable agriculture, capable of astonishing population increase. With the continuance of cereal monoculture black grass is likely to remain a problem indefinitely.

Lolium perenne – *perennial ryegrass*

The genus *Lolium* contains eight species of which *L. perenne* is the most widespread. *L. multiflorum*, Italian ryegrass, and *L. temulentum*, the darnel, with toxic grains, are of agricultural significance. *L. perenne* has a characteristic appearance with deep green shiny foliage and its spikelets set edgeways on to the spike. The species is a common constituent of lawns and playing fields extending, for example, as far south as Cyprus where, on irrigated football pitches, it is the conspicuous survivor in the goal mouth areas (Fig. 9.5). The genus *Lolium* is most closely related to *Festuca* with which it will hybridise in nature. Certain × *Festulolium* hybrids are being developed commercially for forage. *Lolium perenne* concerns many seed companies and the list of recognised varieties amounts to several hundred. A major drawback to this grass is its ability to induce hay fever and the

Fig. 9.5. *Lolium perenne*, here conspicuously visible as darker tufts in a V shape is the species surviving closest to the goal mouth on an irrigated Cyprus football pitch. Other species in the original seed mixture, such as *Cynodon dactylon* and *Poa pratensis*, compete far less well under these circumstances.

effective allergen *Lolp*I amounts to 5% of the soluble pollen protein (Knox and Singh, 1990). Despite this situation, there seems no essential reason in the biology of the grass for *Lolp*I to be produced and the development of low or non-allergen strains would be a most worthwhile enterprise.

Grasses are, with very few exceptions, anemophilous and hay-fever sufferers are aware of how widely grass pollen is dispersed. It is therefore a surprise to find how confined, spatially, the results of cross-pollination are, for which *Lolium perenne* has been studied extensively. Under the conditions of their experiment Copeland and Hardin (1970) showed little outcrossing beyond 6 m. Perhaps unexpectedly, outcrossing was broadly similar up and down wind, a fact attributed to variation in wind direction and turbulence over the three to four week period of anthesis.

Lolium perenne is a well-studied example of the grass self-incompatibility system.

Chloridoideae

Cynodon dactylon – *Bermuda grass*

Cynodon contains about eight species and is closely related to *Chloris* with which it will hybridise. *C. transvaalensis* can be used for lawn grass and *C. aethiopicus* and *C. plectostachyus* are forages. *Cynodon dactylon* is far and away the most conspicuous and successful member of the genus and is ubiquitous through the New and Old World tropics and subtropics. It survives readily, for example, in Italy and there is an outlier population in south-west England. Through its salt glands it will excrete salt onto the leaf surfaces and frequently persists in saline habitats. Figure 9.6 shows a rare variant that occurs throughout the range of the species. Although given varietal status, here it is perhaps best regarded as a subordinate part of *C. dactylon* var. *dactylon* described below.

In a wide-ranging cytotaxonomic study Harlan and de Wet (1969) and Harlan *et al.* (1969) developed a classification they described as 'a compromise between classical morphological concepts of species, the biological concept of species and practical utility'. The authors' difficulties were that, while morphologically discrete groups could be identified, a residual but variable level of crossability could have allowed them all to be sunk in one enormously variable species. In the event, several species were recognised leaving a remainder of *Cynodon dactylon* in three varieties. Var. *aridus* contains diploid ($2n = 18$) forms from very widely distributed regions (South Africa, Tanzania, the Negev and parts of India for example). Var. *coursii* is a tetraploid ($2n = 36$) and is endemic to Madagascar. Var. *dactylon* is the common and ubiquitous weed of the warmer parts of the world. Representatives are tetraploid and, in some cases, highly seed sterile.

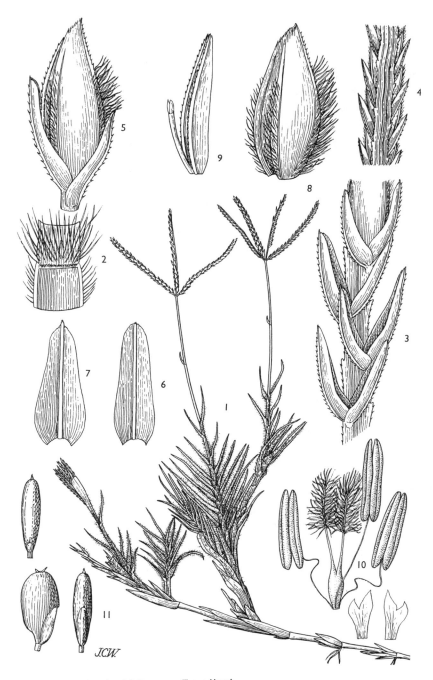

Fig. 9.6. *Cynodon dactylon* (L.) Pers. var. *villosus* Hegel.
1. habit × $\frac{2}{3}$; **2.** ligule × 10.7; **3.** portion of spike after florets have fallen × 25.6; **4.** portion of spike showing rhadus × 12.8; **5.** spikelet × 21.3; **6.** lower glume × 29.8; **7.** upper glume × 29.8; **8.** lemma × 21.3; **9.** palea showing rhachilla produced × 21.3; **10.** flower × 29.8; **11.** grain × 32.

Cynodon dactylon can become conspicuous in overgrazed land (Bosch and Theunissen, 1992) or otherwise extremely disturbed land (Harlan *et al.*, 1969) but, at the same time, offers a useful signal to the conservationist seeking to recolonise a severely degraded landscape, a matter considered in Chapter 10.

Regarding the origin of *C. dactylon* var. *dactylon* the tetraploid, Harlan and de Wet (1969) and Harlan *et al.* (1969) regard varieties *aridus* and *afghanicus* as its diploid progenitors. At the tetraploid level introgression occurs, with minor chromosomal disturbance that itself also promotes significant genetic variation. Occasional seed production from such plants has been sufficient to establish new variants and Harlan and de Wet describe it as 'one of the most dynamic, aggressive and cosmopolitan species in the world'.

At low latitudes *Cynodon dactylon*, in its common and widespread form, creates the impression of an immense resource. A careful search for a range of seed-forming tetraploids seems a self-evident research objective. The situation mirrors that described earlier for *Phragmites australis* in that near inability to produce seed need not impede spectacular ecological success.

Triodia basedowii

The genus *Triodia* is one of a group that includes *Monodia*, *Plectrachne* and *Symplectrodia*, all Australian, that comprise the 'spinifex' grasses of central Australia. *Triodia* has about 40 species of which *T. basedowii* is selected for comment based on Jacobs (1992).

Its growth habit resembles that of a curled up porcupine with grey pointed leaves poking in all directions. As the hummocks develop they may form rings. Seed is produced only in small quantities. Under nursery conditions only about 5% of seed will germinate, a figure that can be raised to 20 or 30% by burning litter over it. Cold vernalisation completely inhibits germination. Seedling survival is improved under shade. Although flowering occurs annually seed set is rare. When seed occurs it falls mostly within the parent hummock with little prospect for survival unless dispersed by ants. Fire is seemingly an essential part of survival in nature and where this removes a substantial part of the parent hummock regeneration can occur from buried meristems and hitherto dormant seeds.

For *T. basedowii* and other spinifex grasses the impression is of heavy reliance on vegetative reproduction giving rise to long-lived plants that withstand and indeed flourish from the effects of fire. They, like *Micraira* discussed earlier, represent a highly specialised evolutionary response to a landscape that may eventually have largely stabilised for millions of years.

Panicoideae

Cenchrus ciliaris – buffel grass

The genus *Cenchrus*, of about 20 species, is recognisably close to *Pennisetum* but convincingly separable from it. *C. ciliaris* (alternatively known as *Pennisetum ciliare*) is a robust variable species found in arid and semi-arid regions of the tropics and subtropics and utilised in appropriate areas as an important forage. Placed in the Palaeotropical kingdom phyto-geographically, the species has been spread widely in the New World and Australia. In the latter it was spread probably by Afghan traders who accidentally or deliberately established it at camel wells about 25 miles apart across the Northern Territory. From these and other introductions it is now a significant component in parts of the Australian grass flora (Chapman, 1992a).

The principal interest of *C. ciliaris* is in its near-obligate apomixis whereby, via apospory, its seeds are maternal copies rather than genuine biparental genetic hybrids. Separate collections of this species readily demonstrate genetic diversity but each accession tends to produce true to type. In cultivation, the rare sexual variants, when they occur, tend to be inherently weak and rather transient. One might expect on this basis, occasional sexual variants in nature, though likely to be found rarely. Crossing an apomict (as male parent) to a sexual one can generate both sexual and apomictic recombinants – the latter conspicuously survives showing hybrid vigour. On this basis the sexual variant has almost the status of a drone honeybee – useful in promoting genetic turnover but otherwise expendable in a population that depends, for its occupancy and spread, upon vegetative reproduction whether by tiller or seed. For a fuller discussion of apomixis see Chapman (1992b).

Hussey *et al.* (1991) reported on a study involving 800 South African accessions. From the most diverse materials it was shown that ovules tended to contain both sexual and apomictic embryo sacs with the latter normally functioning as the result of pseudogamy. In one instance the substitution of self pollen with outcross pollen raised seed set from 36 to 94%.

Cenchrus ciliaris is tetraploid $4x = 2n = 36$. Occasionally instances of pentaploidy $5x = 45$ occur, which is assumed to arise from sexual fusion involving adding nine chromosomes to an unreduced apomictic embryo sac and described as 'B III' fertilisation (Bashaw and Hignight, 1990).

Hussey *et al.* (1991) concluded that the low, but detectable, level of sexuality in their accessions was sufficient to account for the variation observed among morphotypes often widely dispersed and dependent largely upon apomixis.

Tripsacum andersonii – *Guatemala grass*

The New World genus *Tripsacum* has about 12 species and is of particular interest being the only genus capable of hybridising with *Zea*. This fact underlies a controversy between those such as Mangelsdorf (1961) who considered *Tripsacum* to be involved in the evolution of maize and others, de Wet and Harlan (1976) for example, who consider maize genetically independent of it, a subject examined in Chapter 12.

Tripsacum has two well-defined species complexes: *T. fasciculatum, lanceolatum, maizar* and *pilosum* and *T. andersonii, australe, bravum, dactyloides, floridanum, latifolium* and *zopilotense*.

The one selected for comment here is *T. andersonii*, a robust perennial with $2n = 64$. It is largely female sterile, is grown in cultivation, but can occur naturally along streams in the American tropics. Its presumed mode of origin (de Wet *et al.*, 1973) is as shown in Fig. 9.7.

Guatemala grass is therefore a triploid *Tripsacum* to which has been added 10 maize chromosomes, a situation perpetuated by vegetative reproduction. It appears to be, where it occurs naturally, a wild grass to which *Zea* has contributed, possibly on numerous separate instances. Do we have therefore in *T. andersonii* a plant persisting by vegetative reproduction but being repeatedly re-evolved from two well-established seed-forming species?

The foregoing contrasted examples, some perennial, some annual, illustrate the modest, even marginal, role that sexual seed formation can have. It is against this background that annual grasses with total seed dependence and complete holocarpy can be viewed. Barley, rice, rye and wheat, though annual, retain in tiller growth a legacy from perennial

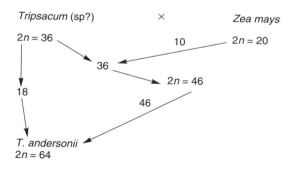

Fig. 9.7. Presumed mode of origin of *Tripsacum andersonii.*

ancestors, a feature now eliminated from uniculm maize. Perennialism of course preserves a given genotype intact and, apart from maize, it is curious that through cleistogamy and thus inbreeding these annual cereals appear to retain another feature of perennials, the perpetuation of a particular genotype.

There remain other curious aspects of grass reproductive biology. One is 'heteroblasty'. In *Aegilops geniculata*, for example, their position in the spikelets may affect the year in which caryopses germinate (Gutterman, 1992). Vernalisation, the cold treatment inducing reproductive growth in winter wheat, represents an interaction between the environment and the *Vrn* allele, 'Vernalisation requirement', of the gene involved. Lastly, conspicuously so in *Festuca vivipara* but widely known elsewhere among grasses such as *Brachiaria decumbens* as shown in Chapter 4, is the curious tendency of the inflorescence to assume a teratogical form converting spikelet to small plantlets and known as 'prolifery' or 'vivipary'. It is appropriate, too, to recall in the present context, those curious instances considered in Chapter 2, where a *Danthonia spicata* infection by *Atkinsonella hypoxylon* can promote a change from chasmogamy to cleistogamy and infection by other endophytic fungi can promote prolifery among *Andropogon*, *Festuca* and *Poa*.

Grass Ecology and Disturbance

10

This chapter is preoccupied with 'disturbance'. It is prompted by an awareness that the bulldozer, power saw, piledriver and concrete mixer presage 'development' and, inevitably it seems, the consequent pollution. A part of the impetus for this chapter is the relentless thud of axes and the plumes of smoke one finds amid tropical hillside forest, and the gullies so often created after heavy rain by discharging water and soil. Who, having witnessed the sudden filling of a dry river bed there can fail to be impressed, not only with the force of uncontrolled nature, but by the human folly, which is its cause? The roar of moving rock and water and the ground trembling beneath one's feat proclaim our mismanagement on an awesome scale.

But, disturbance is not always so arresting. The seemingly benign rhythm of a pump sending water through an irrigation network may signify that we are adding poison to the land driving it inexorably out of production. As the land whitens over, not with harvest but with salt, yet another part of human short-sightedness is exposed. And if this were not enough, consider the immense spoil heaps and lakes of noxious waste that are the residue of ill-planned mine working and the pollution, too, that accompanies the oil industry on the land and in the sea.

Changing Perspectives in Grass Ecology

Whether an ecologist were scoring, for example, the distribution of species in a chalk downland sward or an agronomist studying productivity in pasture, the emphasis has been essentially upon stability and predictability. Sheep nibbled the sward thus retaining its grass population or, if they did not for any length of time, the natural processes of scrub and eventually woodland succession would reassert themselves. Such approaches fashioned in research institutes and universities in temperate countries were

extended and modified at low latitudes. Nowadays, the situation we confront is that permanent pasture and their cycles of regulated feeding have
given place to a devastated environment and to disrupted social structures
seemingly unable to correct matters. Various writers have, in response,
developed for us a heady mixture of overpopulation, climatic change and
political corruption combined with criticisms that Western aid is insufficient, misguided or late. While this alarmist literature has a place in stimulating interest it is not the whole story. It is necessary to find a view that is
balanced but not complacent. Later in this chapter instances of environmental damage will be considered but they are preceded by an instructive
instance from Kenya.

Machacos, Kenya

In 1937 Mayer described the Machacos Reserve as in appalling condition
demonstrating every phase of misuse of land with its inhabitants 'drifting
rapidly to a state of hopeless and miserable poverty and their land to a
parching desert of rocks, stones and sand'.

Almost six decades later the picture has changed radically. Population
has risen but so has agricultural production with soil erosion declining in
what is now a well-treed landscape. An absorbing study by Tiffen *et al.*
(1994) examines both the process of change for the better and analyses the
underlying reasons. The results achieved are undoubtedly those sought
after by much well-intentioned overseas aid. A series of 'then and now'
photographs make the point that what was once a wilderness now has the
appearance of a productive and well-tended garden. What then are the
authors' conclusions?

The Akamba, the people of the region, are considered to be adaptable
and it is not necessary to 'direct' their farmers in regard either to limiting
their families nor how to develop their land. Government support can be
used to increase access to knowledge and to markets and to maintain
infrastructure. Population growth has itself impelled Machacos society to
change and has facilitated the interactions that develop capability for
change. The Akamba are seen as an open society receptive to influences
from various directions such as churches, trade and travel, while retaining
their basic social institutions. In terms of grass biology, two cases are
especially interesting.

Case I

An area of land purchased by a farmer in 1950 was badly denuded, attributed to overgrazing. For ten years the land was closed to grazing. The
farmer collected and planted *Cynodon plectostachyus*, *Aristida kiniensis* and

A. adscensionis (annuals), followed by *Chloris roxburghiana*, *Enteropogon macrostachys* (under trees) and *Heteropogon contortus* (in sandy pockets) associated with *Eragrostis superba* and *E. caespitosa* (perennials). Grazing was resumed but the owner has sustained animal production together with a controlled offtake of wood for charcoal. The farmer has had no technical advice and seeks to maintain the land for subdivision, eventually, to his two sons.

Case II

Kenya achieved independence in 1963. Before that time agricultural policy was, to a degree, imposed and whether it was soundly based or not was hardly likely to draw much sympathy from native Kenyans. This is not to say that potentially useful ideas were invariably ignored. Terracing covered at its peak, in 1958, 42,000 hectares. Thereafter by 1961 it had fallen to 27,000 ha in the absence of compulsion. And yet after independence, terracing began again on a voluntary basis. On the terraces *Panicum maximum* was used to stabilise the banks and *Cenchrus ciliaris*, *Cynodon dactylon* and *Eragrostis superba* were planted to recover grazing land.

Optimism and rationality?

The Machacos example offers practical instances of improvement while pointing to an underlying rationality. A more 'continental' view is that of Timberlake (1985).

> But the direction of Africa's agriculture can no longer come from Washington, Paris and London. It cannot even be run from Lagos, Harare and Nairobi. It must be run largely by the farmers, and by their organisations, with the help and support of the aid agencies and the governments and the research stations. This help and support must come in many small packages, rather than a few big dams and big plantations. And these packages must be delivered with great humility by people who work closely with farmers, people who do not mind when the packages are shaken up and rearranged and perhaps used in a way that the deliverers had not quite intended.

This helps define the problem for ecologists. They are caught in a system where research support is limited and where their objectives can be skewed politically. How far in this context is it possible rationally to dissect the real ecological issues, with grasses in mind particularly?

What is Disturbance?

Long before humankind exerted any influence, a land bridge formed linking North and South America allowing grasses not found in the southern fossil

record to migrate in from the north as described in Chapter 7. Later, in
Pleistocene/Holocene times, stipoid grasses in the north appear to have
been replaced by others such as *Panicum virgatum*, *Bouteloua* spp. and
Buchloë dactyloides migrating northwards.

With the advent of humankind eventually agriculture developed and
with it came the distinction between a crop and a weed.

> Since cultivation allows weeds to survive unaided whereas seed of the new crop
> must be added each season a cynic might conclude that agriculture was more
> suited to weeds than crops.
>
> Chapman and Peat (1992)

There is an interesting sidelight on this. Stebbins (1965) noted the
ingress of non-American weeds in California but drew attention to
indigenous plants which showed weedy propensities, an example being an
Elymus hybrid. The suggestion is that human disturbance confined to the
last two centuries, but on a considerable scale, has thus recently created the
conditions for weed evolution. Two species *E. condensatus* and *E. triticoides*,
both non-weedy, inhabit recognisably contrasted habitats – the latter
consistently more saline. The hybrid between them is easily made
artificially and is seedless. A plant closely resembling this hybrid is found
on disturbed land and, though sterile, spreads efficiently by rhizomes.

So familiar are weeds that we might neglect to ask the question as to
how it is *these* species rather than others demonstrate this degree of
ecological success. Clearly most, but not all, are polyploids, but this is a
feature of about 70% of grasses anyway. Rather more significant perhaps is
the proportion of grasses primarily African in origin, as can be seen from
the next section (see Table 10.1).

Given increased human travel and international trade the number of
weedy species is increasing. For a recent text on tropical grassy weeds, see
Baker and Terry (1991).

'Africanisation' of the New World Tropical Grasslands

Parsons (1970) has shown, in some cases with dates of introduction, when
African grasses have been introduced into the New World. These include
Panicum maximum (guinea) to Jamaica 1741, *Brachiaria mutica* (para)
Angola to Brazil 1823, *Hyparrhenia rufa* (jaragua) Angola to Brazil, 18th
century(?), *Melinis minutiflora* (melado) Angola to Brazil, 18th century(?),
Pennisetum clandestinum (kikiyu) Africa to Brazil 1923, and *Digitaria decu-
mbens* (pangola) Africa to Florida 1935.

These grasses introduced into the Americas have been accidentally or
deliberately spread and progressively replaced the indigenous species. A few

Table 10.1. By taxonomic affinity, the principal grass weeds on a world basis. (Major weeds in bold type.)

Subfamily	Tribe	Subtribe	Genera and species (A = annual, P = perennial)	Chromosome nos.
Arundinoideae	Arundineae	–	*Phragmites australis* P	36, 48, 96
			P. karka P	36
Bambusoideae	Oryzeae	–	*Leersia hexandra* P	48
Chloridoideae	Cynodonteae	Chloridinae	***Cynodon dactylon* P**	36, 40
	Eragrostideae	Eleusininae	**Dactyloctenium aegyptium* A	20, 36, 48
			***Eleusine indica* A**	18, 36
			Leptochloa chinensis A	40
			L. panicea A	?
Panicoideae	Paniceae	Cenchrinae	**Cenchrus echinatus* A	34
			*Pennisetum clandestinum P	36
			*P. pedicellatum P	?
			*P. polystachion P	54
			*P. purpureum P	28, 56
		Digitariinae	**Digitaria abysinnica* P	?
			*D. ciliaris A	?
			***D. sanguinalis* A**	34, 36–48, 54
		Ischaeminae	**Ischaemum rugosum* A	?
		Rottboellinae	***Rottboellia cochinchinensis* A**	36, 40, 60
		Saccharinae	***Imperata cylindrica* P**	20
		Setariinae	*Axonopus compressus* P	40, 60
			*Brachiaria mutica P	?
			***Echinochloa colona* A**	36, 48, 54, 72
			***E. crus-galli* A**	36, 42, 48, 54
			*Panicum maximum P	18, 32, 36, 48
			P. repens P	40
			Paspalum conjugatum P	40
			P. dilatum P	40, 50
			Setaria verticillata A	18, 36
		Sorghinae	**Sorghum halepense* P	20, 40
Pooideae	Aveneae	Aveninae	***Avena fatua* A**	42
	Poeae		*Lolium temulentum* A	14

* Indicates grasses of either African or possibly African origin. Chromosome numbers mostly from Darlington and Wylie (1955).

New World grasses (notably *Digitaria californica*, *Eleusine tristachya*, *Paspalum dilatum* and *P. plicatum*) have arrived in the Old World but their spread in no way matches the success of African grasses in the New World. Parsons (*ibid.*) suggested that African grasses evolved under greater grazing pressure and incorporate therefore an inherent competitiveness (and by implication weediness?).

The African examples Parsons mentioned could be augmented by several others, notably *Chloris gayana* and *Cynodon dactylon*, both selected for mention subsequently.

Two African grasses *Andropogon gayanus* and *Brachiaria humidicola* have been shown in New World savanna to sequester carbon at deep layers in the soil profile, Fisher *et al.* (1994). So extensive are Africanised savannas that the authors consider their scale of carbon fixation to be globally significant.

Deterioration

The increasing human population has created pressure on virtually all aspects of the biosphere. Development of the weed flora and movement of grasses geographically are parts of a trend toward more intensive exploitation. The expansion of cereal culture and overgrazing of pasture amid the fragile ecologies of semi-arid grasslands has led to decline of preferred species and diminution, or even destruction, of the vegetation cover, with consequent erosion. Under irrigation salinization has occurred and, to an unspecified extent, pollution has been an almost inevitable accompaniment. An inland fish farm in Honduras for example can be loaded with pesticide residues and in the rapidly expanding cities of the Third World, grey, foully polluted water discharges through gullies into nearby rivers. Such irresponsible stewardship, seemingly endemic in many parts of the world, finds dramatic expression in the Mediterranean where, in places, human excrement washes on to the beach with all manner of industrial non-degradable rubbish. If this were not indictment enough, a study of that most exquisite animal, the dolphin, found that one in thirty animals had died choking on plastic rubbish (Pearce, 1995). Although not on this scale, abuse of the Mediterranean is hardly new. It has been alleged that in classical times wheat shipped from Libya in vast quantities caused filling, subsequently, of the Rome sewers, which would have discharged into the sea via the River Tiber.

Erosion

Brown and Flavin (1988) estimate that worldwide about 24 billion tonnes of topsoil are being blown or washed off crop land each year. This in turn is partly attributable to the loss of tree cover which again is associated both with lowered water table and proneness to flooding. Like other aspects of deterioration, the problem can be seen in statistical terms or simply by looking at (say) tropical hillsides. A too common sight is of an eroded landscape, almost devoid of topsoil, a sparse tree flora surviving in as yet slightly protected situations, and the occasional clump of *Bambusa vulgaris* holding enough soil together to indicate that all need not be lost.

Salinisation

Salinisation, the accumulation of salt in and on the upper layers of soil, can occur naturally where high temperatures and low rainfall combine to concentrate salt, principally sodium chloride. The process can be induced artificially by installing an irrigation system that makes inadequate provision for disposal of waste water under similar environmental conditions. Poorly managed irrigation systems add to problems of shortage of good land and low productivity. One estimate considers that of 225 million hectares irrigated between 25 and 40% is subject to salt poisoning (IIEDWRI, 1987). In fact it is extremely difficult to make reliable estimates, the principal point being that the problem is both vast and increasing (see Umali, 1993).

It is often advocated that breeding for salt tolerance offers a convenient answer but such thinking avoids awareness of practical difficulties. Gains in salt tolerance could be offset by further rises in salt concentration, and with a greater salt burden against which the plant must work it is progressively more difficult to sustain yield. Rather than misuse the plant breeder in this way, a more rational response would be to challenge the skills of the agronomist and irrigation engineer.

Some Key Features of Grasses Recalled

Inherent in 'disturbance' as the theme of this chapter is the corollary that grasses are able to rectify some, at least, of this disturbance. To demonstrate this it is helpful to recall certain aspects of grasses. In particular they:

1. tend to be primary colonists in many cases;
2. have an elaborate embryo and generous food reserve and can thus make a 'running start';
3. have a relatively 'streamlined' growth habit and can proceed rapidly toward the reproductive phase;
4. are furnished with numerous growth points that aid their competitiveness;
5. are usually drought tolerant in some degree.

The sum of these features is that grasses are opportunists ready to exploit promptly some open habitat especially if the grass available has some additional property, appropriate in the circumstances, e.g. heavy metal or salt tolerance. Grasses, by these means, are on hand to heal landscapes we have torn apart by motorway construction, industrial development, mining, urban sprawl and the requirements of tourism. The Machacos example described earlier can be matched by other instances, each in its way instructive. Not every situation has been remedied and in some cases, given the resources available, the problems can be daunting. What follows is a series of contrasting instances.

Cynodon dactylon *(Bermuda grass) in South Africa*

Earlier, the status of this species was considered and reference made to the treatment proposed by Harlan *et al.* (1969). Among agronomists the persistence of *C. dactylon*, illustrated in the previous chapter, in pasture is seen as 'degeneracy' but the same situation can be viewed by the ecologist as 'tenacity'.

Since this grass exists in many variant forms, could selection among wild stocks reveal especially suitable types for problem areas? Is it possible to define preferred characteristics of a 'reclamation pioneer' and select for it among the diverse forms available?

Fuls and Bosch (1990) collected 140 specimens across the drier parts of South Africa and from these, after growing them out, selected six for further detailed trial. Tests were devised for drought resistance, pH tolerance, biomass production at different fertiliser levels, seed viability and the ability to reproduce vegetatively. Selections were named by reference to their original collection site.

Superior drought tolerance was shown by the Augrabies strain. Across a range of pH from neutral to 2.5 the Delareyville strain displayed outstanding vigour. This same strain showed best responsiveness to nutrient application as measured by biomass, production being on average double that of Augrabies and Clanwilliam strains. As regards reproductive potency, seed viability was invariably low ranging from 0% in Augrabies to 3.5% in Clanwilliam. Vegetatively, Fuls and Bosch (*ibid.*) made a distinction between rhizome and stolon pieces finding the latter thrived more easily. Stolon pieces from Augrabies were superior among the six selections.

The selection of a particular strain for reclamation for a given site depends upon the weighting given to the various characteristics. Fuls and Bosch, while acknowledging the productivity and pH tolerance of Delareyville, stress its low reproductivity. More generally these workers adopt a predictive approach claiming that performance under experimental conditions can be matched to the needs of a particular reclamation site.

An interesting aspect of this work is the approach to seediness. Seedy strains of *C. dactylon* are known and in North America, for example, are commercially available to establish lawns and sports pitches. Fuls and Bosch in fact regard vegetative reproduction as preferable since, compared to seed under drought conditions, the stress period is shorter. In so doing they mirror what *Cynodon* does over much of its range.

There is no doubt that Bermuda grass is immensely variable and that systematic selection can provide competing alternatives for various ecological niches. One conclusion is that *C. dactylon*, so widespread as to attract little attention, should be reconsidered as a valuable resource yet to be fully exploited. This point was made earlier in Chapter 9 and the instance considered here is an instructive example, giving as it does an alternative view of the usefulness of seed production.

Vetiveria zizanioides (*vetiver*)

Vetiveria is an andropogonoid genus allied to *Dichanthium* and *Sorghum*. It grows about 2 m tall, is densely tufted and almost seedless. It is deep rooted, able to withstand both drought and flooding and its tough foliage makes it unpalatable to livestock. It appears not to suffer appreciably from any serious pest or disease. Since some of its tiller shoots emerge from below ground it is, to an extent, fire resistant. The plant's roots yield vetiver oil, an aromatic product. Vetiver grass presents the student of grasses with an interesting situation.

The focus of interest is that a line of slips (tiller pieces) set out along a contour can generate a natural barrier to erosion capable of holding back soil and filtering this from the water runoff. Despite its vigour, the plant is non-weedy and the barrier, once planted, is self-confining. The technique is a proven one and, curiously, has in some but not all parts of the moist tropics long been a feature of agriculture. The present writer first saw it in Jamaica in 1961 where it had already been used for many years.

Given the scale of erosion the problem has been, with vetiver, not to invent a new technique to control erosion but rather to popularise an existing one. The need therefore has been not primarily for research papers but for extension literature, a situation of which the World Bank and the United States National Research Council became aware in about 1985. Through the efforts of many people, notably R.J. Grimshaw, there has been an immense and sustained effort to persuade farmers in the moist tropics of the value of vetiver for terracing land to control erosion. For a popular account see Anon. (1993).

A point of technical interest is how far away from the moist tropics might vetiver be able to control erosion? Attempts have been made to introduce it into semi-arid and even arid regions where it is claimed to reduce wind erosion (Grimshaw, 1992). However, vetiver is hardly frost resistant and certainly unable to tolerate prolonged severe cold since it is, primarily, a tropical plant. In China, for example, the cold-tolerant *Achnatherum splendens* is a feasible alternative and is already widely distributed.

Combating Salinisation

The problem of salt-poisoned land has already been mentioned and the following examples provide contrasting responses. *Puccinellia* is a pooid allied to *Poa*, while *Leptochloa* and *Chloris* are both chloridoids.

Puccinellia chinampoensis

The Hoxhi[1] Corridor, so called because it runs approximately NW/SE between two mountain ranges, is a semi-desert region opening on to the Gobi desert. It is about 1100 km long and between 70 and 500 km wide and is home to more than four million people. The Old Silk Road runs along the Corridor and at one point intersects with the Great Wall of China. Apart from supplying themselves, the inhabitants of Hoxhi provide food for Lanzhou, a city of more than two million people. Evapotranspiration can exceed precipitation at least 15-fold and, in the instance to be discussed, far more than this. The already naturally salty land is vulnerable to further damage from the food needs of the large and growing population. Seasonal temperatures swing from an average of $-10°C$ in January to $21°C$ in July. Day/night oscillation is considerable and in the summer can rise beyond $30°C$. Winter temperatures can fall to $-30°C$ or below. Pressures on agronomists to sustain and even increase productivity are severe.

For Linze, a part of the Hoxhi Corridor having a research station, annual rainfall is about 75 mm and evapotranspiration 3500 mm, about a 47-fold difference. Salinisation, as might be imagined, is considerable and natural vegetation includes *Kalidium*, *Salicornia*, *Salsola* and *Suaeda*, all recognisably halophyte. Salt crusts form readily on the soil surface.

Ren Jizhou *et al.* (1992) have described the following system. Land was cultivated, irrigated and sown with *Puccinellia chinampoensis*. The irrigation water was of high quality derived from an aquifer fed from melted snow on

[1] Pronounced 'Hershi'.

the nearby mountains. Total soil soluble salt was assayed from surface to 45 cm depth over the following seasons and showed an improving trend.

After three years, the sown grass pasture was ploughed and put down to either wheat, linseed, broad bean (notably salt sensitive), millet and sugar beet. *P. chinampoensis* appears to achieve two things. Through its root action there was a considerable improvement in soil structure. Secondly, since its roots substantially exclude salt, this was not carried into the upper layers of soil. The high-quality irrigation water, judiciously applied, both leaches the soil salt and maintains the salt-laden water table below about 50 cm.

Properly managed, the system seems on first acquaintance both elegant and simple. If however it were to be more widely applied there is the problem of disposing of large quantities of salt-laden waste water since this would be both energy intensive and financially almost prohibitive.

Leptochloa (Diplachne) fusca (*kallar grass*)

This chloridoid grass, Fig. 10.1, is widely distributed throughout the tropics, can grow in water-logged saline and sodic soils and excretes salt (sodium chloride) through salt glands on its leaf surfaces. Islam ul Haq and Khan (1971) for Pakistan and Rana and Parkash (1980) for India have advocated the use of this grass for lowering soil salt levels. In Anon. (1984) it is pointed out that, for this procedure to work, salt-crusted foliage must be removed from the site by grazing or harvesting. How useful in fact is *L. fusca*?

The grass is throughout its distribution able to grow on a range of wet soil types to produce foliage palatable to buffalo, cattle, goats and sheep. Although some authorities (Anon., 1984, 1990) advocate planting from slips, some strains are extremely seed efficient and capable, for example, of rapidly achieving weed status in an experimental glasshouse. Since the plant can grow well in unpromising soils but under other conditions is less competitive, Sandhu and Malik (1975) have advocated a 'crop succession' approach to land improvement where kallar grass is recommended as the primary coloniser. Data on the actual saline and sodic tolerance of kallar grass is seemingly scarce although its practical usefulness under these conditions is not in doubt. Given the improvement of soil structure through its root action and effective removal of salt-laden foliage this can be, to some extent, explained. Under experimental conditions it can generate dry matter more effectively at higher levels of root zone conductivity than *Agropyron elongatum* (tall wheat grass), *Cynodon dactylon* (Bermuda grass) and *Hordeum vulgare* (barley) (Sandhu et al., 1981). These authors show that with increasing salinity of the root medium, glycine, betaine and proline increase substantially in the tissue, presumably buffering osmotic stress, and that potassium in the shoot, constant on a tissue water basis,

Fig. 10.1. *Leptochloa fusca* (L), P. Beauv. kallar grass.
1. habit × $\frac{2}{3}$; **2.** spikelet × 10; **3.** lower glume × 10; **4.** upper glume × 10; **5.** lemma × 10; **6.** its palea × 10; **7.** flower × 19; **8.** grain × 10; **9.** ligule × 4.

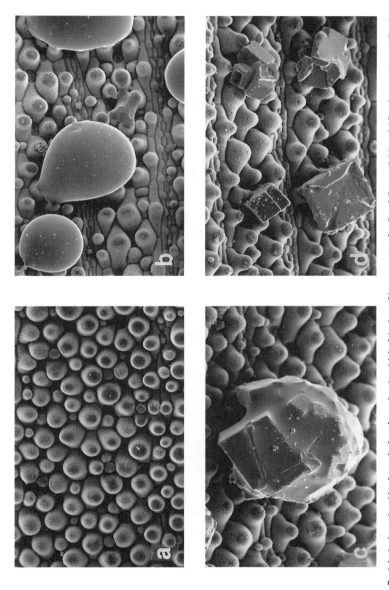

Fig. 10.2. *Dactyloctenium ctenoides* lower leaf surface showing (a) leaf texture, (b) emergence of salt solution, and (c) and (d) subsequent crystallisation × 176. Photo courtesy of Nasruddin Aris.

decreased with dry weight. Another chloridoid grass, *Dactyloctenium ctenoides*, is illustrated in Fig. 10.2, showing the process of salt excretion.

Chloris gayana (*Rhodes grass*)

Chloris is a genus of some 55 species spread through the tropics and mostly perennial. *Chloris virgata*, an annual, can survive extremes of cold by over-wintering as seed. The genus is variously salt tolerant and *Chloris gayana*, Fig. 10.3, though widely dispersed, is highly esteemed as a pasture grass without having achieved a reputation as a serious weed. It is readily esta-blished from seed and through vigorous stolon growth can provide a dense pasture sward and, should the site require it, effectively bind sand.

Given that salinisation is a problem, could its salt tolerance be increased? Malkin and Waisel (1986) subjected *C. gayana* to five generations grown against a background of near-lethal salt concentration and selecting the survivors as the nucleus of the next generation. (Since, in this perennial, the interval to seed formation is short such a programme could be accomplished in less than five years.) Under experimental conditions Malkin and Waisel were able to show that the selected plants could both survive higher salt concentrations and multiple clipping – a requirement of a pasture plant. There is therefore the prospect of selection for salt tolerance but, as was indicated earlier, as an accompaniment to, rather than a substitute for, good irrigation practice.

Arid Grasslands and the Sahel

There is a widespread assumption that the Sahara, for example, is increas-ing in area at the expense of the grasslands at its margin upon which a substantial pastoralism depends. The reality of increasing desertification has been scrutinised recently by Thomas and Middleton (1994). Three factors particularly complicate the situation. Desertified areas and their margins are subject to cycles of climate fluctuations, the proportion of CO_2 in the atmosphere is demonstrably increasing leading to forecasts of a 'greenhouse effect' and the activities of our own species can impoverish the flora of a given area. The reaction to these interacting aspects varies. One is to refine climatic modelling or satellite imaging while another is to survey the present-day situation on the ground. Each has a valid contribution to make but the latter is the most tangible and immediate. This shows that the world's drier grasslands, which have been available to nomadic commu-nities, have undergone vegetational changes although, for the reasons just indicated, the cause is uncertain. Kernick (1990), for example, has shown

Fig. 10.3. *Chloris gayana* Kunth.
1. habit $\times \frac{2}{3}$; **2, 3.** part of spike $\times$ 0.5; **4.** glumes $\times$ 5.3; **5.** lower glume $\times$ 7; **6.** upper glume $\times$ 7; **7.** florets $\times$ 7; **8.** lowest lemma $\times$ 7; **9.** its palea, $\times$ 7; **10.** second and third florets $\times$ 7; **11.** flower $\times$ 7; **12.** grain $\times$ 7; **13.** ligule $\times$ 1.8.

how in Balukistan (Pakistan) *Stipa, Pennisetum* and *Enneapogon*, high grazing value perennials, have given way to *Chrysopogon*. In Rajasthan (India)
Dichanthium, Cenchrus and *Lasiurus*, perennials, have given way to annuals
or vegetation has been eaten out altogether. Given that this induces a
shortage of food, seed collection from grasses such as *Echinochloa* and
Panicum for human consumption might also hinder regeneration of the
grass cover.

At a more extreme level, if land is converted from grazing to arable use
loss of organic matter and decline in water-holding capacity and even
erosion then preclude the ready re-establishment of the grass canopy.

Nowhere perhaps are the trends just outlined more evident than in the
Sahel, the semi-arid margin bordering the southern Sahara. An informative
long-term survey is that of Peyre de Fabrègues (1992). His results show that
from 1962 to 1988/89, a period that included prolonged drought, the
following changes occurred. *Andropogon gayanus, Aristida funiculata, A.
stipoides, Cymbopogon giganteus* and *Eragrostis tremula* declined to zero.
Those species becoming rare or very rare included *Aristida pallida* and
Shoenefeldia gracilis. Some species – for the most part less desirable ones –
have over the same period of time increased in abundance. These include
Cenchrus biflorus, C. prieuri, Dactyloctenium aegyptium and *Tragus racemosus*.

The trend detected in the Sahel repeats that found elsewhere, namely
for perennials to be replaced by less desirable, often annual, plants. A
further complication here is that prolonged drought could have lasted
beyond the period of viability of the replacement perennial species, a point
difficult of which to be sure.

The Channel Tunnel Linking England and France

Engineers have, for 150 years, advocated an undersea link between England
and France. Eventually, in 1994, the Channel Tunnel (the 'Chunnel') system opened, consisting of two railway tunnels with a smaller service tunnel
running parallel between them and linked with them at 100 m intervals.
The whole process involved the removal of about 8,000,000 cubic metres of
chalk marl. Ecological interest is with the means by which disturbance on
such a colossal scale was dealt with in an environmentally sensitive way.
The chalk marl on the French side of the median line was deposited as
slurry in a dry valley behind an earthwork dam at Fonpignon near Sangatte.
The surface area is about 7 hectares and is expected to dry out and be
colonised naturally. The corresponding 4,000,000 cubic metres on the
English side were brought to a site below the cliffs between Dover and
Folkestone to extend an existing 6 ha site to 36, protected by a concrete

retaining wall on the seaward side to create a new piece of land 'Samphire Hoe[2]'.

On this new site an experimental programme was managed by Wye College in conjunction with the Eurotunnel organisation and its contractor Transmanche Link. The primary aim was to stabilise the surface with a vegetation cover which was, in terms of floristic composition, appropriate to the area. Such species would be mostly salt tolerant and typical of chalk downland near the sea. Various seed mixtures were prepared with between 4 and 14 species of wild flower. Different fertiliser applications were used experimentally and the seed mixtures sown with a nurse crop of *Lolium perenne*, a vigorous and highly successful species mentioned elsewhere in this book. *L. perenne* is, however, salt-susceptible with the desired result that, although its effect is to preserve micro-habitats to protect wild flower seedlings, it is the loser in a competitive salt-laden situation and eventually disappears. Indeed much of its 'cover' detected in quadrats is the straw that provides seedling protection (Mitchley and Buckley, 1994, 1995). Mitchley (pers. comm.) considers that to have used *Festuca rubra* would, because of its greater salt tolerance, be likely to have overwhelmed the wild flower seedlings and, for this reason, *F. rubra* was rejected for consideration as a nurse crop.

At the English end of the tunnel, before excavation began, representative species of the wild plant communities were removed, grown elsewhere and eventually returned to rehabilitate the disturbed area (Helliwell *et al.*, 1996).

Ultraviolet Radiation

Chlorofluorocarbons, a group of chemicals used for refrigerants and petrol additives, have been associated with ozone depletion in the upper atmosphere. This in turn has resulted in less efficient filtration of incoming solar radiation, notably ultraviolet B, long known to damage biological systems. The 'ozone hole' was first detected over Antarctica and with it, consequently, diminished productivity in Antarctic phytoplankton (see Smith *et al.*, 1992 for a review). The primary damage is to photosynthesis (photosystem II). Given that ozone depletion was subsequently detected over land masses further north there is then an obvious concern, quite apart from questions of human health, as to whether photosynthesis, particularly in crop plants, might be affected. The two obvious approaches are either to expose plants under defined conditions to known dosages of UV-B radiation or to attempt to associate field performance with altered levels of

[2] 'Hoe' – a promontory or projecting ridge of land. The most famous is Plymouth Hoe where the sailor, Drake, completed a game of bowls before putting to sea to destroy the Spanish Armada in 1588.

incoming natural radiation. The available literature has increased markedly over the past five years and the following examples, for convenience, centre on rice.

In 1991 Teramura *et al.*, under glasshouse conditions, showed that for a wide selection of 16 rice varieties a differential UV-B sensitivity existed. Leaf area, tiller number and biomass were reduced with increased UV-B radiation. One variety, Kurkaruppan, showed, surprisingly, a significant increase in biomass under these circumstances.

Under glasshouse conditions, Barnes *et al.* (1993) demonstrated among 22 varieties of mostly *indica* rice less sensitivity to increased UV-B radiation in upland United States varieties commonly planted and more in modern high-yielding lowland types developed in the Philippines. Although yield diminished, the more obvious changes were those in shoot morphology.

He *et al.* (1993), in a comparison of pea (*Pisum sativum*) with two *indica* rice varieties, showed that the cereals were less damaged by corresponding doses of UV-B. In a further study, He *et al.* (1994) demonstrated the relative effectiveness of 'hardening' induced by low levels of UV-B as a protection against more severe doses. Again the rices performed better but a further point of interest was in their contrasted responses. 'Er Bai Ai' relies on the accumulation of UV-B-absorbing compounds while 'Lemont' depends, apparently, on the increased synthesis of protective water-soluble enzymes.

It would hardly be surprising if there were different levels of sensitivity to UV-B either among different species or among different varieties within a species or that resistance might be differently organised. While selection for UV-B tolerance might happen anyway in the course of normal field trials, it is unclear how far it should, as yet, be deliberately sought.

These contrasted examples indicate something of the scale of problems we face and the means by which grasses help us to do something about them. Ten thousand years ago humankind confronted the difficulty of finding sufficient food. In that situation grasses were the remedy to hand and became the major constituents of agriculture.

Natural Cataclysms

Not all large-scale disturbance is attributable to our own species. Among the most spectacular disturbances, and ones over which we have no control, are volcanic eruptions.

Vesuvius is a well-known volcano near Naples in Italy. In AD 79 its effluvia engulfed the towns of Herculaneum, Oplontis, Pompeii and Stabiae. It has a history of repeated eruption, explosively so in 1858, 1906 and 1944. Careful examination of the volcano and its adjacent lava flows means that for these dates we have examples of advanced, advancing and recent succession. Surfaces persisting from 1858 are relatively well vegetated, those of 1906 slightly less so and those of 1944, in some cases, even now, barely colonised, although taken together they are very informative ecologically. Among 25 pioneer flowering plants on the 1944 lava flows were eight grasses: *Aira caryophyllea*, *Briza maxima*, *Bromus rigida*, *B. tectorum*, *Cynosurus echinatus*, *Dactylis glomerata*, *Holcus lanatus* and *Vulpia myuros* (Mazzolini and Ricciadi, 1993). On the looser materials of the cone *Bromus erectus* was found in the shelter of *Rumex scutatus* (*ibid.*). Ricciardi (pers. comm.) indicates that particularly on the looser materials of the cone, more so than the lava, grasses are important among the first colonists. Here in 1988/89 were found among 35 flowering plant species (many of them woody and representing later succession), the grasses *Agropyron repens*, *Aira caryophyllea*, *Bromus tectorum* and *Vulpia myuros* (Mazzoleni *et al.*, 1989).

Human depredation on the volcano is evident. Ricciardi *et al.* (1986) published an account of the Vesuvius flora with 610 species but with 293 taxa previously known to have occurred there now missing probably due to extensive afforestation and disturbance from the, almost certainly ill-advised, extension of Naples on to the slopes of the volcano. Time will tell.

Vulnerable Grasses

The thrust of the present chapter has been to show that, for the most part, however regrettable ecological disturbances have been, grasses are normally able to exploit and stabilise the situation. This is not invariably the case and it is appropriate to recall here those rare bamboos inhabiting Bahia and other tropical forests referred to in Chapter 7. How far does their vulnerability recall an earlier stage in grass evolution before the emergence of the conspicuously pioneering habit that characterises many of the Poaceae?

Domestication   11

Finally, everyone must judge for himself whether it is more probable that the several forms of wheat, barley, rye and oats are descended from between ten and fifteen species, most of which are unknown or extinct, or whether they are descended from between four and eight species that may have either closely resembled our present cultivated forms or have been so widely different as to escape identification.

Darwin (1868)

One view of grasses is that they are only or chiefly interesting because they provide our cereals and our principal forages. From such a standpoint almost all other grasses could appear marginal and inconsequential. An alternative view, and that of this book, is of cereals and forages best seen within a wider context as *some* rather than *the* members of the Poaceae. Viewed in this way, cereals are a late development among a minority of grasses whose future is closely, even inextricably, linked with a somewhat wayward primate, *Homo sapiens*. Such grasses have the advantage for us that they provide special and closely studied instances of accelerated evolution. Indeed, it is instructive that a family which had existed for perhaps 60 or 70 million years, within the last 10,000 years found the means to react so successfully to the challenges and demands presented by our own species. None the less, the response has been 'patchy'. Most grasses were not drawn into the cereal mode. Those that did mostly came in early and those grasses important now have been so for most of the domesticated period. It is appropriate therefore to regard the 'cereal minority' as a group of atypical grasses that, even so, have useful lessons to teach.

Origins of Domestication

The eruption of Vesuvius in AD 79 mentioned earlier, which overwhelmed Pompeii and other nearby towns, preserved all manner of artefacts and, associated with them, the remains of cultivated plants – not least various members of Poaceae. These include, from Pompeii, *Panicum miliaceum* and *Setaria italica*, and from Herculaneum, *Avena sativa, Hordeum sativum* (six-rowed) and *Triticum dicoccum* (Meyer, 1980). AD 79 is relatively late on the agricultural time scale and these species differ little from their present-day equivalents. To resolve, if possible, the problem presented by Darwin there is need to seek the earliest indications of agriculture. Were we able to find them would the crop remains differ in instructive ways from their modern counterparts?

The oldest materials so far discovered are probably those at Jericho dating from about 10,000 years ago. Subsequently, in order of appearance, there were the agricultures of South China (about 8500 BP), North China (7500 BP) and Mexico (about 5000 BP). 'Jericho' used in this sense relates to an archaeological site near the modern town of that name and excavated by Kenyon (1978), Figs 11.1 and 11.2. Not surprisingly, a good idea, namely agriculture, was adopted fairly rapidly elsewhere in the Near East including Jarmo in Kurdistan, excavated by Braidwood (1960).

As Kenyon points out, the Jericho site is interesting because, at so early a date for agriculture, already the cultivated forms of *Hordeum sativum*

Fig. 11.1. A bowl of stored grain recovered from Jericho during the 1953 excavations.

Fig. 11.2. A stone quern or hand mill (top) used for grinding corn, recovered from the 1955 Jericho excavations.

(two-row) and *Triticum dicoccum* were in place, at an appreciable distance from where their wild relatives occurred. An important feature at Jericho is a perennial water spring. Kenyon concluded that even at so early a date irrigation too would have been in place and with it sufficient social organisation to control its use. The evidence is of a settled society producing a range of artefacts. In this sense archaeological evidence for cereal origins runs into the sands of time. We cannot see a biological transition from wild to cultivated forms and thus have to find another way to solve the problem Darwin set us. This is by exploring genetic differences, experimentally, between cultivated plants and their presumed wild relatives. Curiously, a biological transition *is* available in the archaeological evidence for maize and makes an interesting contrast to the situation described by Kenyon for barley and emmer wheat. For a recent treatment of agricultural origins see Smith (1994). Since the separation between wild and cultivated grasses is, in evolutionary terms, so recent, the two groups have diverged only minimally and the breeder therefore can utilise wild grasses in support of cultivated breeding programmes. The principal differences are, under domestication, the substitution of non-shattering rachises in place of those which fragment, free-threshing caryopses instead of those invested with persistent chaff, and some moderation of dormancy.

Such a view, given that experts differ on details, provides the basis of what students are taught about evolution under domestication, the origins of agriculture and what nowadays is called 'classical' or 'conventional' plant breeding. It is, in this chapter, neither appropriate nor necessary to challenge this current orthodoxy nor to repeat what is conveniently set out in numerous other texts. Comment, presently, is reserved for other, more open, issues.

From Cultivation to Domestication

In what European settlers dismissively described as 'firestick' agriculture the Australian aborigines had long before developed a system of controlled burning of patches of bush. Griffin (1992) has examined the ecology it represents and shown that it has a convincing rationale for maintaining vegetational stability. By this means an aboriginal technique imposed on the environment, if not cultivation, is at least a degree of management directed at sustaining a preferred flora (and fauna). Within this, seed collection was practised that required a subtle awareness of a living community. Latz (pers. comm.) described how *Yakirra*, a relative of *Panicum* but possessing oil bodies (elaiosomes), is collected by ants. Both bandicoots (a marsupial mammal) and aboriginal gatherers collect seeds from the ants' nests. The aborigines then put the caryopses and associated debris in a 'coolamon', a shallow wooden bowl, and by 'yandying', a form of winnowing, separate the caryopses (Fig. 11.3). Awareness of such a system must surely challenge the idea that seed gathering is a random, ill-considered process preceding the development of any identifiable technology. Elsewhere Harlan (1989) has stressed the close awareness of hunter-gatherers of the ecology of their situation.

This, in turn, leads to what supposition we make about the origins of agriculture. A distinction is made between 'conscious' and 'unconscious' selection of non-shattering, free-threshing types. Did the Neolithic farmer find, to his surprise, after some decades of cultivation that these new types had come to predominate? Alternatively, did his already finely tuned awareness suggest to him that some types of inflorescence were preferable to others? We cannot know. We ought however to do two things. The first is to note with admiration the *durability* of the Neolithic farmers' choices.

Fig. 11.3. An Australian aboriginee 'coolamon', an ornamented shallow wooden bowl used for winnowing chaff from edible seeds. This example is about 25 cm in length.

What, thousands of years ago provided the staples, such as barley, oats, rye and wheat, do so today. If such farmers were *this* perceptive about the many grasses from which they had to choose perhaps we should less readily assume that non-shattering, free-threshing types came to predominate only accidentally.

The second thing we could do is critically reconsider the idea that 'man' domesticated cereals. Given the enthusiasm with which men depart for the hunt, the unpredictability and indeed the apparently diminishing success of the enterprise in former times, it seems likely to me that nursing mothers, confined at home, with time to think and children to feed, might have dreamed of a more predictable, perhaps more humdrum, way of meeting need. Should we not say rather that, in all probability, women domesticated cereals? Even today in many countries they share the brunt of sowing, harvesting and sifting, shown in Fig. 11.4. Did they themselves invent the burden, much of which they continue to bear?

'Domestication' redefined?

'Domestication' formerly meant being brought within the household economy. Darwin recognised that this redirects selection and different biotypes come to prominence. Breeding takes the process a stage further by applying

Fig. 11.4. A Chinese woman hand-sifting wheat grains to remove the remaining grit and chaff after it has been threshed and winnowed.

selection to both parents rather than just the female. If one adds to this the other ingredients of modern technology, domestication could be redefined as 'bringing within the household of science'.

On this basis, many grasses, in addition to the cereals, are domesticated, including those used, for example, for forage which have been subjected to controlled breeding. It could be argued that even this is an insufficiently comprehensive view and that it is necessary nowadays to regard virtually everything as a domesticate once we begin to control the ecology of some environment. There are some, including this author, uneasy with this approach. An alternative response, recalling the aboriginal burning policy, would be to recognise three levels of increasing control: management, cultivation and domestication. The first, to the grass biologist, would cover, for example, practices leading to deflected climaxes, the second implies deliberate planting and the third deals with significantly changed biotypes as the result of genetic selection. To apply these categories

Fig. 11.5. *Saccharum officinarum* (sugar cane) showing a readily fragmented branchlet of the inflorescence detaching spikelets in the manner of a wild grass, × 15.

loosely then has the advantage that one can recognise degrees of domestication. *Digitaria iburua* (fonio), for example, is a West African cereal so far subjected to almost no scientific improvement. At the other extreme are maize, rice and wheat, which together drive much of modern plant science. Since fonio is to an extent non-shattering and free-threshing it is domesticated. It is however useful to recognise, at this point, that *Saccharum officinarum* (sugar cane), a grass well understood genetically, and, nowadays, subject to highly sophisticated breeding programmes, none the less retains the wild grass characteristics of a fragile inflorescence and chaff-associated caryopses, shown in Fig. 11.5. The reason is that selection has been applied, not to the reproductive, but to the vegetative organisation of the plant. Even so it would be unrealistic to regard sugar cane as anything other than highly domesticated.

The Neo-Darwinian Synthesis and Cereal Improvement

Although Darwin elucidated natural and artificial selection he was unaware of his contemporary Mendel who demonstrated 'particulate' inheritance. Neither Darwin nor Mendel were aware of chromosomes and their significance. Only when, in the twentieth century the chromosome theory of inheritance, various aspects of population genetics and Mendelism were combined with various insights of Darwin, was it possible to synthesise 'neo-Darwinism'. Such a synthesis underlies much of present-day plant breeding. It might, equally accurately, be described as 'applied cytogenetics'.

How far could 'classical cytogenetics' be expected to sustain cereal improvement and how soon should this be abandoned in favour of newer techniques?

'Modernist' Genetics

In 1983 Herrera-Estrella and separately Zambryski reported transformation of a higher plant using *Agrobacterium tumefaciens*. This major discovery stimulated 'plant biotechnology' in which, not surprisingly, cereal companies were caught up. Unlike the situation in dicotyledons, no convincing examples of transformation were found for grasses which, in this sense, came to be seen as refractory. This in turn stimulated an interest in alternative systems, notably electroporation and microprojectiles for DNA delivery. The results so far, though not negligible, are small compared with what *Agrobacterium* technology has achieved with dicotyledons. One can, therefore, take 1983 as the beginning of a 'modernist' approach and that time preceding this as 'classical'. The distinction is a useful one so long as it is not too rigidly applied since much present-day grass research is a blend of classical and modernist (or molecular) techniques.

Nowadays, the radial loop model for the eukaryote chromosome is widely accepted but fitting DNA sequences into it remains elusive and indeed among cereal geneticists is a relatively long-term objective. None the less, closely detailed linkage maps of the major cereals are now available and will be considered later.

'Classical Cytogenetics'

Avdulov (1931), already referred to in Chapter 6, provided a major contribution with data for chromosome size and number of many genera for grasses.

In 1932 C.D. Darlington's book *Recent Advances in Cytology* appeared, an immensely influential work across the whole of biology which made a profound impact on the study of grasses. More than 20 genera of grasses were included to cover such topics as hybrid origins of polyploids, genetic control of chromosome size, haploid parthenogenesis, 'B' chromosomes and the elaborate analysis of meiosis. In a real sense this book wrote much of the agenda for cytogenetics for the next half century. Shortly after its publication the effects of colchicine were recognised – primarily in constricting chromosomes at metaphase of mitosis and in disrupting spindle formation to simplify the creation of polyploids (reviewed by Eigsti and Dustin, 1955). Darlington, earlier, had included several pooid grasses because they were accessible, economically important and possessed reasonably large chromosomes that could be conveniently examined. The involvement of American scientists and the wider range of latitude represented had stimulated an interest in *Zea*, notably through the very accomplished work of McClintock, discussed subsequently. As cytogenetics became a more worldwide interest, increasingly a larger number of grass genera were studied.

Before the midpoint between 1931 and 1983 the progress of grass cytogenetics can be conveniently gauged from G.L. Stebbins' book *Variation and Evolution in Plants* (1950), a wide-ranging conspectus of cytology, genetics, plant geography and taxonomy. Much of Darlington's agenda had been worked through although the emphasis on temperate genera remained.

In 1953, Watson and Crick announced the structure of DNA and redirected the focus of biology. At the same time anxieties about feeding the world's population were emphasised. In 1969 Darlington gathered together his assessment but by now objectives had begun fundamentally to shift. The real interest for us at this point is to identify amid classical cytology the harbingers of the modernist approach.

'Pointers'

1. Gene to chromosome relationship

The smaller the region of a chromosome that can be identified, whose effects can be pinpointed in the phenotype, the closer we are to a particular gene. If with improving biochemistry we can analyse such a region with increasing accuracy, eventually it should be possible to present it as a series of nucleotides in which we can distinguish active and non-active components. Among the most convenient chromosomes for such work are polytenes (laterally replicated many times) found in insect salivary glands. Polytene chromosomes occur in plants, notably in antipodal nuclei, but lack characteristic banding patterns that facilitate gene location. Pachytene of meiosis can provide an acceptable substitute and that of *Zea mays* is especially suitable since the knobbed appearance of each chromosome is characteristic and can be fairly readily prepared for examination.

Creighton and McClintock (1931) bred a strain of maize heterozygous for two changes to chromosome 9. These were, on one homologue a terminal knob and an interchange (translocation) from chromosome 8. The same homologue contained the dominant *C* for coloured aleurone and the recessive *wx* for waxy endosperm. The other homologue lacked both terminal knob and the interchange. It possessed *c* for colourless aleurone and *Wx* for starchy endosperm.

In the offspring 'knobbed' always occurred with coloured aleurone and 'knobless' with colourless aleurone. Waxy types (deriving *wx* from both parents) showed the presence of the interchange in 11 out of 13 progeny. Two individuals showed a crossover between the *Wx.wx* locus and the interchange. This work was the first to demonstrate for any organism the equivalence between genetic and cytological crossing over. Later Rhoades (1945) established that *Dt*, a gene influencing colour dotting in aleurone, was also located in the knob region of chromosome 9.

In 1987 Da-leng Shen *et al.* hybridised a labelled probe 10.8 kb long for 'waxy' to pachytene chromosomes of *Zea mays*. Labelling was concentrated around a single site in the distinctive knob region of chromosome 9. *In situ* hybridisation provided support for a conclusion long known from classical genetics, namely 'waxy' as a single locus (single copy) gene.

2. Exceptional gene behaviour

An axiom of genetics is that chromosomes carry genes, linearly arranged, which can be presented as a linkage map. An inversion or translocation changes the order in which genes occur. Additionally, a mutation at a given point increases the alleles available for the particular gene residing there. These comments would apply to any eukaryote.

Of special interest, therefore, are any real or apparent exceptions. If, for example, nature presented us with a model system of how genes shifted among chromosomes could we develop this to insert genes that particularly interested us?

In 1950 McClintock described the presence of *Ac*, an 'activator' gene seemingly causing mutations in genes on other chromosomes. It operated through *Ds* (on another chromosome) where genes close to the latter mutated so as to produce a variegated phenotype. It was surprising that such genes as *Ac* and *Ds* appeared occasionally to move from one chromosome to another. Interest in this early work resurfaced more recently. Freeling (1984), using more modern terminology, asked whether 'transposons' could be used deliberately as integration molecules. If successful it could perhaps provide for monocotyledons (or at least maize) a matching system to that involving *Agrobacterium* in dicotyledons.

3. Differential staining

It has been known for many years that a metaphase chromosome need not stain uniformly but can show, laterally, a banded appearance. More recently, it has been recognised that if a cytological stain were sufficiently discriminating the resulting differential staining could indicate fundamental differences in the composition of the chromatin, namely differences in the DNA. (Additionally, if otherwise similar chromosomes have dissimilar banding patterns they can be individually characterised.) Giemsa stain is suitable for these purposes. For details of the 'C-banding' technique see Gill and Kimber (1974a,b). In wheat, most bands occur near the centromere although some nearer the ends also show up. In rye, the ends themselves are conspicuously stained. As a result, the chromosome contributions of each species can be identified in wheat × rye hybrids.

Although the dark-staining bands are called 'heterochromatin' or 'constitutive heterochromatin', they do not contain only one kind of DNA although much of it comprises satellite DNA consisting of many-fold repeated sequences that are not transcribed. Such DNA is not, moreover, entirely confined to such bands but also scattered to a lesser extent throughout the chromosomes. And sprinkled among it is that minute fraction, less than 1% of the total, the coding DNA which is transcribed and is at the basis of gene action.

4.   DNA amplification

To the bulk of DNA in the chromosome we can ascribe no function. Detailed analysis shows various meaningless sequences repeated almost endlessly, changed perhaps over generations but apparently to no purpose. The matter has been carefully explored in wheat, rye and their close relatives, and results from rye (diploid and therefore more convenient) are presented by Bedbrook *et al.* (1980) and Appels (1983).

Telomeric heterochromatin contains between 12 and 18% of the total nuclear DNA. When analysed, four repeated sequence families having respectively 480, 610, 120 and 680 base pair repeat lengths were found. The 680 and 610 types are confined to the telomeres. The 480 is mostly telomeric and the 120 is found at both telomeric and interstitial sites (Bedbrook *et al., ibid.*).

A related species *Secale sylvestre* is stained far less intensively by Giemsa and in this species the 480, 610 and 630 repeat families are absent or nearly so. By contrast the 120 is present in about equal amounts in *S. cereale* and *S. sylvestre*. The 120 base pair sequence occurs not only in rye but among wheat species. As with species divergence (assessed on other criteria) so sequence divergence is also greater (Bedbrook *et al., ibid.*). While, therefore, sequence divergence can be used as one indicator of species relationship, the role of such DNA for the organism remains a mystery.

5.   RFLPs (restriction fragment length polymorphisms)

Certain enzymes will fragment DNA to relatively small pieces, a so-called 'library'. Such enzymes act upon particular and characteristic conjunctions of bases. If the fragments are then sequenced it can become apparent that one fragment contains within it information found in others. By such a laborious process it should be possible eventually, in stepwise fashion, to sequence even the entire DNA content of the human genome. Recent developments have made it possible to automate the process removing some of the drudgery and delay.

6.   RAPDs (randomly amplified polymorphic DNAs)

RFLP technology, being relatively slow, has prompted a search for alternatives. If, say, a short length of DNA is amplified many times by the 'polymerase chain reaction' (PCR), the recognition of such lengths of DNA from a chromosome provides another kind of locus and one that eventually could become part of a linkage map.

7.   Isoenzymes (isozymes)

Flowering plants are rich in enzymes and frequently a situation exists where several enzymes are available to complete a particular reaction. An organism can thus possess for example one of each as a pair (a heterozygote) or be homozygous for one or other. An important point is that, using electrophoresis, each homozygote and the heterozygote are all recognisable. Since within one plant variants of several such enzymes can be present, any plant can have a distinctive 'fingerprint' or 'signature'.

The following examples explore the results of combining 'classical' with 'modernist' genetics.

Mixed linkage maps

In classical genetics, the emphasis was upon morphological 'markers', genes recognisable by their phenotypic effect which, in a cereal, could include stature (tall or dwarf), grain colour, disease or pest resistance, endosperm texture and many others. Given relatively few chromosomes and numerous markers, *Hordeum* ($2n = 14$) and *Zea* ($2n = 20$) became early examples of detailed linkage maps. (To achieve such results, it was of course quite unnecessary to know anything of the DNA sequence of any gene involved.)

With the advent of DNA technology it became possible eventually to combine the results of classical and new approaches of which an example is that for barley (Kleinhofs *et al.*, 1993). Here the seven chromosomes of barley are presented as a series of linkage maps in which 295 loci deriving from RFLP, RAPD, isozyme, morphological, disease resistance and other studies are combined. (It is noteworthy that the paper has 22 contributors representing 12 scientific departments in various parts of the United States.) A similar map has been described by Devos and Gale (1993) for hexaploid wheat with comparable data for the seven chromosomes of each of the A, B and D genomes. These authors also present a comparison for the A genome and chromosomes of rye, *Secale cereale*.

It will be self-evident that in these examples there is a detailed understanding of gene order and linkage relationships.

To concentrate for the moment on wheat, classical cytogenetics offers the breeder a formidable array of options. He or she can, for example, by using monosomics substitute from one variety to another any one of the 21 chromosomes unchanged. Through the use of bridging species such characters as disease resistance can be brought from relatively remote sources genetically and incorporated into, say, bread wheat. Despite misgivings on theoretical grounds some years ago, yields continue to rise and, in less than a century, through hybridisation with rye, wheat has given rise to a new group of cereals the *Triticosecale* (Triticales). For a recent review of wheat cytogenetics see for example Lagudah and Appels (1992).

There remains a tantalising problem. The DNA studies implicated so far are primarily *analytical* rather than *ameliorative*. They help describe the present content of chromosomes rather than how they could be permanently altered. As indicated earlier, since *Agrobacterium* techniques are inapplicable to cereals, what evidence is there that others based on a modernist approach, through molecular biology, can work instead? The key requirements are that a gene can (a) be introduced, (b) will integrate itself into the genetic system, (c) can be expressed, and (d) will be capable of inheritance preferably in a conventional Mendelian manner. Three examples provide informative contrasts.

1. Rice transformation by electroporation

Datta *et al.* (1990) describe a protocol where protoplasts derived from *indica* rice pollen are treated with DNA carrying hygromycin resistance in the presence of polyethylene glycol (PEG). The cells derived from such protoplasts would normally be haploid. These authors note the presence of spontaneous fusions and any fusion involving two protoplasts would be expected to create a diploid situation. From such treated protoplasts, cells, embryonic callus and eventually mature seed-setting (and therefore diploid) rice plants were obtained. Seedlings raised in the next generation remained hygromycin resistant. Whether or not the added gene could be inherited in a Mendelian manner was not resolved in this publication.

2. Barley transformation by microprojectiles

Wan and Lemaux (1994) described a method of barley transformation using, not protoplasts but immature embryos dissected from caryopses. Either whole or longitudinally dissected embryos or callus derived from scutellum tissue were bombarded with gold particles on to which DNA had been adsorbed. Plants were obtained which were shown to be both self-fertile and herbicide resistant and from these it was found in some cases that resistance could be passed on to the next generation. Again, these authors did not examine experimentally possible Mendelian patterns of inheritance.

Interest in both electroporation and microprojectiles (biolistics) is now considerable and Table 11.1 indicates some of the many results becoming available.

If at some point following the original transformed generation, differences appear among the progeny – some carrying and expressing the added gene and some not – this of itself should not be construed as

Table 11.1. Examples of transformation among various grass genera 1993–1994.
(a) Electroporation

Organism	Transformation	Reference
Agrostis alba	Resistance to amino-glucoside antibiotics achieved by addition of neomycin transferase II	Asano and Ugaki (1994)
Saccharum officinarum	Incorporation of β- glucuronidase	Arencibia *et al.* (1994)

(b) Microprojectiles

Organism	Transformation	Reference
Agrostis palustris	Embryogenic callus regenerated to yield β-glucuronidase activity	Zhong *et al.* (1993)
Hordeum × *Triticum* (Tritordeum)	Immature inflorescence tissue explants yielded stable incorporation of β-glucuronidase and neomycin transferase	Barcelo *et al.* (1994)
Triticum	Scutellar tissue treated and regenerated as plants with β-glucuronidase activity	Nehra *et al.* (1994)

'Mendelian'. Only when transformed plants have been systematically crossed with untransformed ones and progenies repeatedly conforming to Mendelian ratios obtained could such a claim be made justifiably.

3. Apomixis in *Pennisetum*

The genus *Pennisetum*, containing about 80 species, has representatives with $n = 9$ and $n = 7$, the former perhaps more primitive. Those species on $n = 9$ are both polyploid and perennial and in some degree apomictic. Those species on $n = 7$ are sexual and include polyploids that are perennial and several annual diploids, notably *P. glaucum* (*P. americanum*), pearl or bulrush millet.

Pearl millet is conspicuously protogynous. Elite stocks therefore are readily contaminated by related pollen creating a problem for maintenance of varietal identity. If, therefore, obligate apomixis were transferred from a wild polyploid to pearl millet, a uniformity might be imposed making this cereal easier to maintain as recognisable varieties, an idea explored by

Dujardin and Hanna (1988). A simplified outline of their hybridisation programme is given below and illustrated in Fig. 11.6.

Backcrossing to pearl millet is expected progressively to (1) lower chromosome number and (2) restore the pearl millet phenotype, while (3) allowing recurrent selection for apomixis derived from *P. squamulatum*.

Two difficulties remain. The first is the rather unpredictable inheritance of apomixis and the second is that introduction of germplasm from *P. squamulatum* also tends to introduce male sterility. Is there an alternative approach? One option appears to be to involve a third species, *P. purpureum* (Napier grass), with which *P. glaucum* has long been known to be cross-compatible (see Fig. 11.7). The second approach is 'Trispecific' involving three species of which *P. glaucum* is common to both double cross parents.

Again, 93% of some 2000 progeny were male sterile, but among the male fertile plants good pollen was available. Eventually one derivative was found to be both apomictic and with 73% stainable pollen. This particular part of the work substantially increased the number of seedlings available thus allowing the breeder more room for choice.

Work over about a decade has resulted in backcross progenies with 27, 28 and 29 chromosome plants. One plant produced 89% maternal progeny although the goal is 100% (obligate) apomictic progeny. Some backcross plants had progeny with only aposporous embryo sacs thus recalling the *P. squamulatum* parent (Hanna *et al.*, 1993).

How far can this work be translated into molecular terms? By using some 48 RFLP probes, it was found that seven hybridised to a restriction

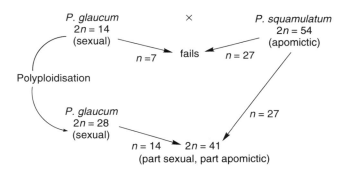

Fig. 11.6. Failed and successful crosses between $x = 7$ and $x = 9$ *Pennisetum* species.

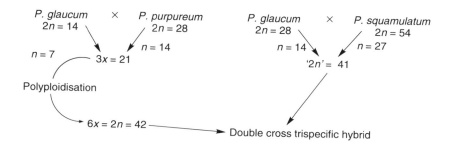

Fig. 11.7. Broadening the genetic base by adding material from *P. purpureum* based on $x = 7$ and closer to *P. glaucum* than *P. squamulatum* (such a process provides new germ plasm and extends the opportunities for the breeder).

fragment in an apomictic 29 chromosome plant and to the *P. squamulatum* parent. Of these seven, one was of particular interest since among the pedigree of 29 chromosome plant it hybridised only with apomictic versions. There is the prospect therefore of selecting apomicts in the seedling stage rather than having to wait until flowering (Ozias-Akins *et al.*, 1993).

Modern Cereal Breeding

The improvement of maize, rice and wheat still largely depends upon so-called classical or traditional methods and seems likely to do so for the foreseeable future. Increasingly the techniques of electroporation and microprojectile bombardment will probably make a contribution. There is however a complication. While much of rice research is in the public domain that for maize and wheat is seemingly more confined by commercial considerations.

Genetic conservation

Although dozens of plants qualify as cereals in various stages of domestication the number that have real economic significance in a global sense is small. Maize, wheat and rice dominate the situation and to these can be added barley, bulrush millet, and sorghum. In a subordinate role, oats and rye can be included. Beyond these are the minor millets of local significance, such as t'ef in Ethiopia and fonio in West Africa.

A generation ago it was recognised that the greater the dependence on a given cereal the more important it was to preserve its genetic diversity and that of its close relatives. Put simply, the wider the genetic base available the less vulnerable a crop would be to some disease or other, a vulnerability through failed harvest which could extend to those of us who depended on it.

There now exists, internationally, a network of 'gene bank'/'genetic' depositories and a degree of administrative and academic support at various levels of sophistication. It is probably fair to say that the material in store now substantially exceeds what breeders will be able to utilise but that is preferable when the challenge to him or her is so unpredictable. To take one quite realistic example – if the climate of a region were to change recognisably but the food preferences of the inhabitants remained similar the breeder would have to respond to both the physiological adaptedness of the plant and the almost certain intrusion of either new pests and diseases or the presence of familiar ones in more insistent forms of attack.

Gene Pools

Although, primarily, genetic conservation concerned the major crops, it now embraces an immense range of plant types. Some rare palm or orchid which needs to be multiplied so that it can avoid extinction becomes a challenge to tissue culture. Agrostologists for their part have collected a range of grasses from arid and semi-arid environments (Prendergast et al., 1992). Apart from efficient sampling of natural populations, interest concerns the de-velopment of a technology for optimal storage conditions and, in the case of cereal relatives, their precise genetic relationship to the cultivated species.

Many years ago Harlan and de Wet (1971) put forward the very practical concept of primary, secondary and tertiary gene pools – defined as follows:

1.   Primary gene pool – the cultivated species and those with which it can be readily crossed giving normal meiosis at F_1 and the production of F_2 seeds.
2.   Secondary gene pool – includes species from which genes can be transferred to the cultivated species with some difficulty. F_1 plants may be weak and the recovery of segregates problematic.
3.   Tertiary gene pool – includes species taxonomically so remote from the cultivated species as to require 'special techniques' such as embryo rescue in *in vitro* culture.

Depending on the finesse with which a particular breeder works, a given cross could be more or less feasible and allocation to a particular gene pool relative to a given cultivated species depends on individual judgement. In the *Pennisetum* example previously quoted involving the transfer of apomixis one would probably allocate *P. squamulatum* to the secondary gene pool relative to *P. glaucum*.

For an informed assessment of declining variation in crop plants and practical responses to that situation see Hawkes (1983). Part of the value of that book is the comprehensive and broadly favourable treatment it gives to the work of Vavilov, a Russian plant geographer who, through a vast collecting programme, identified for cultivated plants 'centres of diversity' from which he drew less readily accepted conclusions that these could also be 'centres of origin' for cultivated plants. For an alternative view of Vavilov's work see Harris (1990).

'Elginism'

Genetic conservation has been politicised and an analogy with the 'Elgin marbles' is helpful. In 1687 the Turkish army demolished much of the Parthenon in Athens, having used it as an ammunition store. More than a century later, Lord Elgin, an English nobleman, retrieved sculptures from the ruins of the decorative frieze and brought them to England. Eventually, they have been set out under virtually ideal conditions in the British Museum in London. Who owns them? Are they Greek, British or, as a unique human achievement, the property of humankind? Should the United Kingdom retain or return them to Greece? Many Greeks would prefer this. Others, with a less direct involvement, and noting the traffic pollution in Athens might, for the present, prefer to leave them in London.

There are similarities with germplasm. Material has been collected, in the past, without permission. To whom does it belong – a collector with the imagination to realise its importance and the zeal to gather it or to the country in which it happened to be growing as part of its natural resources? Fortunately, the analogy with the Elgin marbles breaks down since seeds can be multiplied and a portion returned to its country of origin. Nowadays an international agreement, The Convention on Biological Diversity, a product of the 1992 Rio de Janeiro Earth Summit conference now ratified by about 100 countries, governs seed collection and importation into other countries.

Gene 'storage' versus gene 'manipulation'

For the purposes of discussion do we now have to invent a 'quaternary gene pool'? If a bacterial gene, for example, is introduced into a crop plant it surely far exceeds the taxonomic distance envisaged by Harlan and de Wet (1971) separating, say, wheat and even its most distant relatives in the tertiary gene pool. And if a gene were not from any living organism but synthesised in a laboratory before being inserted into a living organism where it could reproduce and be inherited, do we propose a fifth 'quinary' level of remoteness? At this point gene collection itself could become a redundant concept as what we need we would increasingly invent. So far, however, this is for the future as much of cereal improvement continues to follow traditional lines.

The Cereal Domestication Achievement So Far

Cereals are annuals and the net result of cultivation is commitment, ultimately, of much of the plant's biomass to seed production. Most of this is available for consumption and a small proportion is reserved to create next season's crop. Humankind has thus available a self-renewing store. The role of modern science is to fine-tune a Neolithic idea first thought of some ten thousand or so years ago.

The Future

Evans (1993) presented an interpretation of maize evolution showing yield trends over five millennia rising to a value for the USA of about 6.5 tonnes per hectare. Because there has been a three-fold increase in yield per hectare over the last 80 years what can we assume about the next 80 – that yields will stabilise, continue to improve at the same rate or even accelerate? The longer the time interval over which we seek to predict, the greater the uncertainty and in this case population increase, global warming, CO_2 enrichment, rising sea level with loss of arable land, and presently inhospitable land at high latitudes becoming more congenial to agriculture combine to multiply our uncertainties. In view of this will scientists redouble their efforts on behalf of maize or expend effort in other directions?

The same questions might be asked about any of the major cereals and, as Evans (*ibid.*) remarks, if methane, a by-product of both cattle production and rice paddy biosis, is produced in sufficient quantities as to compound global warming should we not modify our agriculture accordingly? In a different area of concern, is there merit in a scientific effort to make our cereal grasses nitrogen-fixing?

Nitrogen-fixing Cereals – a Holy Grail?

A strand of current thinking is that tropical soils, low in nitrogen, exist within economies for which the cost of applying artificial fertiliser is prohibitive. The traditional response has been to advocate the use of legumes that fix nitrogen, following almost automatic infection by *Rhizobium* or related bacteria. An extension of such thinking is to explore whether cereals might be modified so as to fix nitrogen in a similar manner. The following questions might be asked.

1. Will *Rhizobium* infect grasses?
2. If not, could it be 'made' to do so?
3. If a grass were infected would it form nodules and do these structures permit nitrogen fixation?
4. Can there be shown overall advantages in doing so?
5. Or is a yield penalty, unacceptably high, incurred?
6. If *Rhizobium* and its allies cannot effectively infect grasses then could other organisms usefully do so?

Not least among the problems of trying to answer these questions is the following likely situation. Suppose under highly specific and unusual laboratory conditions a grass is shown to be infected by some target symbiont and suppose too that the system tested for nitrogen fixation is shown to be at the very limits of experimental resolution. Does the investigator under these circumstances claim either that nitrogen fixation by a grass 'appears to be possible' or that it 'is unproven'? In the first case does his enthusiasm justify trying to refine the methodology or, in the second, is his objectivity a safeguard against the imprudent use of further resources? To carry conviction that grasses *do* fix nitrogen two things are necessary. Firstly, that the conditions under which they do so are 'not so unusual as all that' and secondly the amounts of nitrogen fixed have to be 'not negligible'. It is against this background that the various claims and counter claims should be judged.

Interest further increased after 1961 when Dobereiner described nitrogen-fixing bacteria in the rhizosphere of sugar cane. A series of investigations eventually established the idea of nitrogen-fixing bacteria in the rhizosphere and *on* the roots of several tropical grasses. There thus developed the notion of 'associative' or exogenous nitrogen fixation. For a review see Dobereiner and Day (1975). There are also more recent reports; for example Cavalcante and Dobereiner in 1988 announced their discovery of a nitrogen-fixing system involving *Acetobacter nitrocaptans* functioning *within* the roots, stems and leaves of sugar cane which was thus 'endogenous'. In 1992 Dobereiner *et al.* claimed the presence of *Herbaspirillum* species in roots of several grass seeds, sorghum and napier grass, and in roots, stems and leaves of rice and sugar. The matter remains contentious and Lee *et al.* (1994) for example were unable to demonstrate agronomically significant associative nitrogen fixation in pearl millet and sorghum.

In a separate line of investigation Cocking *et al.* (1994) claimed that maize, rice and wheat formed nodules containing rhizobia and were able to fix nitrogen from the air. The work is at only an early stage of investigation and no field data are available.

How might a plant breeder react? Suppose, eventually, that it were shown unequivocally that sugar cane with its abundant carbohydrate did in fact support appreciable nitrogen fixation. One possibility, no more adventurous than attempting to nodulate cereals, would be to search for 'grain-forming sugar'. Although sugar is a cultivated grass, selection, as was described earlier in this chapter, has been applied to the vegetative parts, the inflorescence being essentially that of a shattering wild-type grass – a situation which helped complicate early breeding studies. Given the remarkable extent to which the genetics of sugar cane is understood, it could be feasible to search for non-shattering types and to shift some of the immense photosynthetic potential from vegetative to reproductive activity. Provided enhanced grain production did not significantly inhibit subsequent transmission of the endogenous diazotroph this too might be a route to a nitrogen-fixing cereal.

Whatever the future may hold, it seems likely that maize, rice and wheat will continue to monopolise the interests of cereal scientists and, because of its importance and its inherent interest, it is with one of these, maize, that this book concludes.

Maize 12

... the ancestor of corn was corn and not one of its two American relatives, teosinte or *Tripsacum*.

Mangelsdorf (1958)

The progenitor of modern corn is probably the wild grass known as teosinte.

Beadle (1980)

... the maize ear is the transformed, feminised, and condensed central spike of the teosinte (male) tassel that terminates the primary lateral branches.

Iltis (1983)

This chapter begins from the assumption, subsequently to be explored, that maize is an 'unusual' even 'strange' grass. Since this plant is so much a feature of world agriculture and in many places the preferred crop, it may well be that maize is seen as an everyday, commonplace or staple item and its seasonal cycle as integral to the farming year. To suggest therefore that this familiar and convenient grass is somewhat out of the ordinary could evoke surprise. In fact, any sense of surprise should be directed toward the plant itself.

The 'Familiarity' with Maize

Not only is maize found in the small garden or across hundreds of hectares, it is also endlessly photogenic. It figures in the American musical *Oklahoma*. Colombians, for example, will model Nativity figures out of the husks of corn ears, but maize has long been a source of artistic inspiration. Hundreds of years before the arrival of Columbus in the New World, the maize ear and its grains inspired numerous pottery designs.

199

The 'Convenience' of Maize

Not only is maize familiar, but it is also 'convenient', as can be appreciated by comparing it, under peasant farming conditions, with wheat. Wheat is cut by sickle, requiring at some point thereafter the separation of straw and chaff. The grain adhering to the rachis and invested by chaff has to be 'threshed'. If, as is the case in some countries, it is done by a mule pulling a heavy corrugated roller over a circle of the harvested crop, it will be necessary at some point to separate the mule excreta from the chaff and grain. The next process, 'winnowing', involves, in a light breeze, throwing shovelfuls of grain up in the air so that most of the chaff is blown away. Finally, to remove the last remaining debris of grit, chaff and small pebbles, it is necessary to sift out the unwanted material and pick over the wheat grains by hand as illustrated in the previous chapter (Fig. 11.4).

By contrast, the maize harvest is very straightforward. The ear is cut from the plant, the husks pulled off and the grains rubbed off the cob. It is quick, simple and, compared with the wheat example quoted, far more hygienic.

Additionally, while at an earlier stage of maturity, maize can be eaten as sweetcorn, the close association of chaff and small grains makes this impracticable for wheat. Wheat, or at least bread wheat, retains its place in world agriculture by its adaptability to a range of environments, its productivity and because of the preference so many people have for eating leavened bread. Modern mechanisation has both obscured for us the traditional inconvenience of bread wheat and simplified the harvest of other types such as durum.

Finally, the convenience of maize can be underlined by its co-cultivation with a climbing legume such as *Phaseolus* in the Amerindian farming tradition. The stout maize stem provides a convenient support for the twining legume and the distance between the maize plant 'hills' makes harvesting both corn ears and bean pods convenient. And, the maize hill receives its quota of nitrogen from the legume to benefit whatever is next planted after tillage.

Maize as a Grass

The features that make maize so familiar and convenient are precisely those that make it unusual among grasses. Figure 12.1 illustrates a conventional modern maize plant. The following list points out contrasts between maize and grasses generally in each case with the maize features placed last as, presumably, those more recently derived.

Fig. 12.1. *Zea mays* L.

1. habit $\times \frac{1}{16}$ showing terminal male inflorescence and lateral female inflorescence; **2.** part of male inflorescence $\times \frac{2}{3}$; **3.** fruiting female inflorescence, enclosed at the base in sheathing bracts $\times \frac{1}{2}$.

1. Perennial versus annual habit.
2. Several stems rather than a single one.
3. Inflorescence normally a panicle rather than a spike.
4. Hermaphrodite rather than a monoecious plant body.
5. Hermaphrodite instead of a unisexual inflorescence.
6. Terminal rather than 'intercalary' (female) inflorescences.
7. Shattering versus non-shattering (female) inflorescences.
8. More commonly, several inflorescences per plant instead of two or three in some peasant farming type maize and finally the single ear in the most modern plant form.
9. Grains invested by glumes, lemma and palea as contrasted with naked (and enlarged) grains.
10. The mature inflorescence free of the subtending flag leaf and not surrounded by several modified leaf bases (the husks).
11. The arrangement of grains in panicles rather than highly condensed into multiple (vertical) grain rows resulting from polystichy.
12. Independent of, rather than dependent upon, humankind for survival.

Viewed botanically, maize is not so much unusual as astonishing. Moreover, this in turn has prompted questions about maize origins that, even after decades of enquiry, yield controversial answers and provide some of the impetus for this chapter. Even so, it is possible, at this stage, to outline some of the essentials of maize evolution.

Evolution introduced

In general terms we can say that a wild grass subject to human selection within the context of Amerindian agriculture evolved into modern maize. By implication, less than 10,000 years ago nothing closely resembling it existed, and it is the most remarkable example we have of change under domestication. Since change has been so far reaching, it is consequently more difficult both to identify the progenitor and to pinpoint the progressive steps by which change occurred. Was there only one wild grass involved? Do we look for very few 'catastrophic' changes or numerous small steps? What is the precise nature of Amerindian farming and the potential of a wild grass that in combination could yield so extraordinary a result? Before offering answers to these problems, it is necessary to view the plant as a working unit.

Inflorescences

The inflorescences of maize, terminal tassel and intercalary ear, to a casual observer are regular and consistent features but closer examination shows that not only is each 'unstable' but also manifests instability in ways that are botanically significant.

Genetic instability

Maize breeders are aware of 'off-types', genetic variants that have little place in mainstream commercial maize breeding but which have had considerable interest for those concerned with the evolutionary origins of corn. Such variants provided Mangelsdorf and co-workers (Mangelsdorf, 1958, 1961; Mangelsdorf *et al.*, 1964) with the means to create (and perhaps in so doing, re-create) 'primitive' maize.

The modern maize breeder works within defined constraints imposed by growing season, the mechanical requirements of plant architecture, available farm machinery and the needs of the market-place. It is understandable therefore that, almost unconciously, the breeder accepts a particular phenotype consisting of a single stem, a single ear medianly placed, 'naked' grains firmly anchored to the cob and conforming to a relatively narrow range of caryopsis size, colour and content.

Maize need not be like this. Any of the above features can be varied widely by recourse to genetic alternatives. If, for example, we sought to produce a plant having several ears, their grains being covered by chaff, we could readily do so. It would be entirely possible to produce forms with grains produced in the tassels or to produce forms with grey, red or brown caryopses or having their wall more than one colour.

Physiological instability

The phenotype referred to earlier with the median ear and terminal tassel is retained provided the maize plant is growing within the area to which it is physiologically adapted. To grow tropical maize in a temperate area or vice versa readily induces a changed appearance and implies that orderly development of the familiar plant habit depends upon the operation of physiological triggers. Given the genetic make-up of maize and all manner of concealed and subtle variation it is hardly surprising that plant response to changed conditions yields diverse and even conflicting results. In what follows, data are described from a small selection of the vast literature that exists on this topic.

Experiments fall broadly into three groups, those where a plant is transferred from one environment to another for its entire life cycle, those where at predetermined intervals during its life transfer from one environment to another occurs, and those where for some tissue rather than the whole plant, the effects of growth regulators are explored. The first approach reveals relatively gross differences, the second helps identify periods of sensitivity to some environmental stimulus or other and the third seeks to attribute specific functions to known chemicals.

McClelland (1928) for maize in Puerto Rico (and thus tropically adapted to short days and high temperature) increased day length to 15 h gave increased plant height, delayed silking and decreased grain yield. Ragland *et al.* (1966) with temperate-adapted maize found that supplementary lighting increased the number of kernel initials but did not result in more filled grains. Hunter *et al.* (1974) studied the combined effects of temperature and photoperiod using temperate and tropical-adapted varieties. As daylength was increased so did vegetative growth and with a delay in tassel initiation. As temperature increased, the magnitude of the photoperiodic effect declined. In order of responsiveness 'Gaspe Flint', an early maturing variety, was least responsive, two Ontario-adapted hybrids were intermediate and 'Pioneer × 306' a tropically adapted hybrid, most responsive. Kiniry *et al.* (1983) studying 20 varieties found that photoperiod sensitivity varied among them significantly. What is incidentally interesting about this work is that least sensitivity occurred among early 'lines' (open-pollinated types) while the medium and late maturing hybrid varieties were more so where, of course, breeders would tend to concentrate their efforts so as to maximise yield through longer life cycle.

The foregoing work applies to treatments given throughout the plant's life. Need a plant however be equally responsive to a given stimulus throughout its life or might sensitivity wax and wane through time?

Moss and Heslop-Harrison (1960) using 'Golden Bantam', a temperate-adapted maize, applied 30 long days treatment and thereafter short days changing from 18 to 8 h days – a treatment which caused partial feminisation of the tassel. Struik (1982) began plants in long days and after a period switched them to short days. The converse treatments were also applied. Among the results it was found that for a short period only long days could influence the number of tassel branch primordia. Struik used two temperature regimes d18°C/n12°C and d30°C/n24°C. By moving from 12 to 20 h daylength at the higher temperature regime the time interval between anthesis and silking rose to 12 days. By moving from 20 to 12 h in the lower temperature regime the anthesis to silking interval was reduced to less than a day but with the order reversed, silking being slightly ahead. Sex reversal could also occur; for example, with a change from 12 to 20 h at the higher temperature some female flowers became male. One option that might occur to the reader would be to space a series of changes over smaller intervals or multiply the temperature regimes. This however might be described as 'experimental overdrive' since the point has already been made that temperature, photoperiod and genotype interact to prompt the diversity of response of which maize is capable. To what extent can this be explained biochemically?

The most direct answer to this question so far has been achieved via the technique of culturing pieces of young inflorescence *in vitro*. It will be recalled that both staminate and pistillate inflorescences are *potentially* hermaphrodite but tend preferentially to express one or other sex. For the production of viable pollen Polowick and Greyson (1982, 1984, 1985), Paraddy and Greyson (1985) and Paraddy *et al.* (1987) progressively refined the *in vitro* process.

In a matching series of papers Bommineni and Greyson (1987, 1990a,b) and Bommineni (1990) not only improved progressively conditions for ear culture but demonstrated, too, the masculinising effects of kinetin + GA_3 (gibberellin). For a review see Chapman (1995).

One curious and thought-provoking example comes from plant pathology. *Oscinella frit*, the frit fly female, can oviposit into the tassel male flowers. These hypertrophy and generate galls remarkably similar in appearance to maize caryopses and from which the next insect generation emerges. Clearly, some kind of hormonal stimulus provided by the insect opens up an otherwise closed developmental pathway thus exploiting for the insect's own use the potentially feminine aspect of the tassel.

A résumé

It will be evident that the ear and tassel of maize each harbour the potentialities of the other. These can be activated genetically or physiologically even when, for example, as in the last case, this is mediated by an insect.

In all of these foregoing examples, the experimentalists hardly created novel forms but in various ways activated alternative ontogenies.

A Co-ordinated Approach

Dellaporta and Calderon-Urrea (1994) recently attempted to begin harmonising genetic and biochemical data so as to resolve the developmental pathways of monoecy in maize. To do this they cited various genes and considered their interaction with hormones and environmental factors. The essence of their approach is as follows.

1. Maize flowers are initially bisexual (hermaphrodite) and proceed to unisexuality by selective arrest and abortion of the inappropriate sex organs.
2. Mutants are known with specific effects on floral structure. These include *dwarf* (*d*) mutants (*d1, d2, d3* and *d5*), *anther ear* (*an1*) and *D8*.

All mutants are andromonoecious (perfect flowers in the primary ear

Fig. 12.2. Linnaeus' specimen of maize which is interesting because it manifests tassel seeding. The caryopses and silks are clearly visible in the proximal portion of this terminal inflorescence, × 1.

florets and staminate flowers in the tassel and secondary ear florets).

d1, *d2*, *d3* and *d5* block the gibberellin biosynthetic pathway.

3. Environmental treatments that feminise tassels such as short daylength and low light, have also been correlated with increase in gibberellin-like substances in the tassel.

4. Other mutants such as *silkless* (*sk1*) prevent the formation of pistils in the ear.

5. Masculinising genes are known where their mutants such as *tassel seed* (*ts1* and *ts2*) producing pistillate flowers in the tassel.

These authors regard tassel seed dominants as antagonistic to the feminising effects of gibberellins for example. By combining *tassel seed 2* and *dwarf 1*, it is possible to create hermaphrodite florets in both the ear and the tassel. In summary therefore, the manipulation of known mutants can have predictable consequences for maize inflorescences and their components. Alongside this is an emerging awareness of the role of growth regulators. Finally gene action could be, therefore, understood in terms of its interaction with various hormone levels.

A Taxonomic Perspective

Taxonomists have little difficulty recognising the close relatives of maize even though, as indicated earlier, maize itself is peculiar. A key factor for taxonomists is the separation of the sexes. Watson and Dallwitz (1992) accept the tribe 'Maydeae' containing *Tripsacum*, *Zea* and *Euchlaena* (teosinte) and five other genera, all Old World.

If the Old World genera are omitted and *Euchlaena* merged with *Zea* then *Tripsacum* and *Zea* comprise the subtribe Tripsacinae of Clayton and Renvoize (1986). Doebley and Iltis (1980) and Iltis and Doebley (1980) have proposed a more detailed alternative that emphasises the close relationship between maize and teosinte.

Regardless of which taxonomic treatment is adopted, those investigators most closely concerned with understanding the evolution of cultivated maize have confined virtually all the discussion to three entities, namely *Zea mays*, teosinte (however regarded taxonomically) and the genus *Tripsacum*.

Against this genetic, physiological and taxonomic background it is now feasible to consider contrasting opinions about maize evolution.

Competing theories

At the head of this chapter are three contrasted quotations that encapsulate different views of maize evolution. Each view has significant scientific support and none deserves cursory dismissal. Figure 12.3 summarises points of contrast among them.

1. The Mangelsdorf hypothesis

The view advocated by Mangelsdorf in a series of publications is that wild corn, now extinct, was a multi-tillered annual with inflorescences each having proximal female and distal male spikelets. The results of selection have been ear enlargement with the acquisition of multiple grains and a more complete separation of the sexes to give a median ear and terminal tassel. Tiller number has been reduced to create the modern uniculm and the other distinctive properties of maize outlined earlier.

An essential point of the early Mangelsdorf approach is that maize owes significant changes to the introgression of genes from *Tripsacum*, and that teosinte is a late product of maize evolution – a weed co-evolved with it and not pre-dating the origin of maize itself.

Wild maize is held to have become extinct through competition for habitats within the emerging new crop and because of cross-pollination from the cultivar producing hybrids that diluted the characteristics of the wild grass. Finally, it was thought conceivable that goats, introduced after Columbus, would have completed the demise of this already rare endemic. Early references include Mangelsdorf (1958, 1961) and Mangelsdorf *et al.* (1964). Before reading the Beadle hypothesis or that of Iltis, the reader is advised to see the work of Collins (1919). This paper is useful in the extent

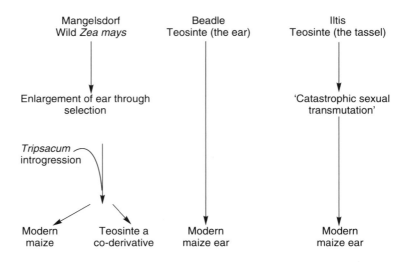

Fig. 12.3. Three alternative theories of maize origins

to which it recalls ideas already in circulation and is remarkable in foreshadowing developments some 60 years later.

2. The Beadle hypothesis

Teosinte and maize can be readily hybridised, a feature hardly surprising since, Beadle contends, the former is the wild species out of which the latter arose through selection under domestication. When the two species are hybridised it is possible to segregate a range of intermediates. By arranging these as a graded series between teosinte and maize one gains some idea of the steps that could have led from the wild to the cultivated plant.

Neither *Tripsacum* nor any other taxon is required, it was considered, to account for the properties of modern maize. See Beadle (1980).

3. The Iltis hypothesis

Two major difficulties in accounting for maize evolution are its soft-shelled grains and its polystichy – the possession in the ear of numerous vertical grain rows.

Iltis sought to meet both these difficulties by proposing that the *tassel*, the male inflorescence of teosinte, which incidentally lacks a central spike, underwent a process of fasciation whereby the lateral branches became fused and also converted from male to female expression as they were displaced to the region of feminising influence. Such a process required what Iltis called 'catastrophic sexual transmutation' which would explain how intermediates were neither formed nor required by theory. The principal reference is Iltis (1983) and recalls a view first canvassed some 90 years earlier (Anon., 1893).

Among the three hypotheses, whichever is chosen, only part of the requirement is to explain the present form of the cultivar so widely current. It is also necessary to explain the extraordinary diversity of maize, its physiological responsiveness to environmental change and the ability of its inbred genetic lines to manifest, when crossed, such remarkable hybrid vigour among the progeny.

Competing theories considered

It is easier to deal first with Iltis' idea of catastrophic sexual transmutation, CST, since it comprises perhaps a single dramatic change. In meeting the problems of explaining how a hard-shelled caryopsis gave way to a soft-shelled one and the origin of polystichy it creates other problems. If the male inflorescence were removed to the region of feminising influence, thus displacing the female inflorescence, why should it not, in becoming feminised, adopt the existing female characteristics it already possessed? Secondly, if a tassel lacking a central spike by fasciation creates polystichy, from where did the modern maize tassel arise that *does* have a central spike?

Neither Beadle nor Iltis saw any role for *Tripsacum* in maize evolution. It is therefore curious that in *Tripsacum* a mutant recalling CST occurred. Dewald *et al.* (1987) described a mutant form of *T. dactyloides* where (a) the proximal single-flowered male spikelets are replaced by double-flowered female spikelets and (b) the staminate spikelets in the distal portion are replaced by perfect (hermaphrodite) ones. The inflorescence in its mutant form is thus gynomonoecious. The condition is due to a single major gene recessive allele and the authors consider that 'the feminising gene found in *T. dactyloides* provides tenuous support for the "catastrophic sexual transmutation theory of Iltis"' (*ibid.*)

Perhaps, too, it recalls Vavilov's idea of homologous variation where genetic possibilities in one species can be matched among its relatives (Vavilov, 1920, in translation by Chester 1949/50) and it is hardly surprising if they share a common ancestor.

Setting the Iltis hypothesis aside, is it possible to apply some tests that could decide in favour of either Beadle or Mangelsdorf?

Teosinte

For Beadle, teosinte must pre-date maize while for Mangelsdorf teosinte is not original but derivative. If therefore the earliest preserved specimens of teosinte are demonstrably older than the earliest maize of which we have evidence it establishes, not that maize did, but that it *could have* originated from teosinte.

It is a feature of Amerindian agriculture that the debris from this activity was accumulated in various caves in which there had been human habitation. Moreover, given the dry atmosphere of Central America such cave deposit material would dehydrate rapidly rather than decay. There is thus the fortunate circumstance that thick layers of such debris preserve a record of Amerindian farming. Even more fortunate, such material is stratified, the oldest layers being at the bottom and the most recent at the top. A vertical section can therefore represent a time sequence that, through carbon dating, can be shown to encompass many hundreds of years and recognisable episodes or culture phases of habitation. Typically, a cave is periodically abandoned and, after a lapse of perhaps several decades reoccupied by a different cultural tradition. Through such profiles it can be shown that maize remnants, fragments of ears particularly in the lowest earliest layers differ appreciably from those added more recently.

Such 'prehistoric' plant deposits have acquired a crucial significance in the search for maize ancestors, a point re-examined later. For the present, do such deposits show early evidence of teosinte?

MacNeish (1985), having presented evidence of excavations from ten regions from Arizona to Guatemala, involving about 25,000 archaeological corn cobs and some 10,000 assorted fragments and representing a time span of up to 8000 years BC, concluded that maize invariably precedes teosinte. This worker does include teosinte material and even evidence of its hybridisation with maize. Seemingly though, these occur invariably in the later, higher levels.

There is no evidence of comparable magnitude known to the present writer supporting the opposite case for teosinte preceding maize. One record, Lorenzo and Gonzales (1970) claims an ancient provenance for teosinte comparable with the oldest maize. The claim is discounted by MacNeish (*ibid.*). Several workers (Collins, 1919; Beadle, 1980; Doebley, quoted by Raloff, 1993; Galinat, 1995) have demonstrated that maize and teosinte will hybridise and from the segregation patterns have drawn the reasonable conclusion that only relatively few genes with their alternative alleles control the difference between maize and teosinte. The further conclusion that this somehow establishes, therefore, the primacy or likely primacy of teosinte, is more of an assertion and appears to ignore the voluminous archaeological evidence reviewed by MacNeish (*ibid.*).

Maize itself

It is necessary at this point to disentangle two sets of ideas. One is that teosinte, as the possible ancestor or early participant or as a later derivative, had some involvement in the evolution of maize. This concern is with the status of teosinte which, as the foregoing section indicates, is controversial.

The second set of ideas is far less contentious and merely concerns the observed increase in size of the maize ear through time. The focus is not on causes but effects and what the archaeological evidence unquestionably establishes is the quite remarkable increase in ear size and grain row number over the time since maize was first domesticated.

Productivity in maize can be partitioned into several factors and these, collectively, describe rather than explain the changes in the evolving ear.

The evolving ear

The starting point for any structural interpretation of changes in the maize ear must be the recognition that it is an andropogonoid grass. Typically, such grasses have a 'spikelet pair' arrangement consisting of one sessile and one pedicellate spikelet. Although in modern maize the ear shows a massively congested version of this arrangement, it is abundantly clear on the tassel spikelets. Here the sessile and pedicellate spikelet pairs are both committed to pollen production and, functionally, only anthers are present.

Logically, one could then assume that an uncomplicated arrangement of sessile and pedicellate spikelet pair characterised the female portion of whatever grass preceded maize domestication, a feature it would share with other andropogonoid grasses.[1] If the pedicellate spikelet became grain-bearing, there is then a grain pair.

At this point there is a choice which it is as well to recognise. One could derive a series of events in the proposed evolutionary history of the maize ear that have their basis in hybrids with teosinte. These have two disadvantages, neither necessarily crucial. The first is the complicating feature provided by the existence of the 'cupule', an arrangement whereby the grain is partly enclosed in the case of teosinte. The evolving maize ear on this basis requires the elimination or modification of the cupule. Secondly, if the archaeological evidence were conclusive, teosinte would not have been available. There are thus advantages in keeping open the possibility that the maize ear evolved independently of teosinte and this is the alternative choice open to us.

There is different and further consideration about whether to involve teosinte. It alone shares with maize the clear separation into intercalary female and terminal male inflorescences – a clear and compelling reason, it might seem, to regard it as a likely, or even the best, candidate for a maize progenitor. Again, its absence in the earliest archaeological record must make us cautious and we should not rule out the possibility of its separated inflorescence being a result of its co-derivation with maize.

How far, then, is it possible to proceed without invoking teosinte in accounting for the maize ear in its present form? In the section that follows, reference to teosinte hybrids indicates the use of teosinte as a 'tester' stock to explore the genetic system and does not necessarily implicate it as founder material.

A 'Minimalist' approach to changes in the maize ear

The following events, not necessarily in this order, attempt to define the fewest possible steps needed to change from a wild grass inflorescence to a modern maize ear.

1.   Reactivation of the pedicellate spikelet
At the level of the spikelet pair two grains are produced where, by analogy with other andropogonoids, one might be expected. On morpho-

---

[1] By analogy with *Sorghum* (see Chapter 4), for example, the pedicellate spikelet might have no function normally although seedy variants can occur there. In the case of *Sorghum*, however, it should be recalled that the functional floret of the sessile spikelet is hermaphrodite rather than female.

logical grounds reactivating of the pedicellate spikelet so that it might become grain-bearing is the most obvious explanation. (In maize × teosinte hybrids $Pd:pd$, paired versus single functional spikelets operate.) A subsidiary development would have been a shortening of the pedicel to make the two grains equally disposed about a node.

2. Increase in vertical grain rows
One vertical row of paired grains was duplicated and then either reduplicated (eight grain rows) or added to piecemeal (six grain rows), and so on until modern grain row numbers were achieved. (In maize × teosinte hybrids many-row versus two-row spikes are segregating alternatives.)

3. Change to non-shattering inflorescence
A change from a shattering type must have occurred, perhaps reinforced by the development of an axis supporting multiple-grain rows.

4. Increased spikelet pair number
The length of the axis and with it the number of spikelet pairs per grain row would have increased.

5. Development of naked (non-tunicate) grains
There was a diminution in glume, lemma and palea length relative to the enlarged caryopsis so as to create 'naked' grains. The Tu allele substituted by its recessive has the pleiotropic effect of reinforcing feminisation.

6. Husk development
If the ear were produced on a side branch local condensation resulting in compacted nodes would have congested the bases of successive leaves creating the husked condition.

7. Efficiency and feminisation
The most 'efficient' ears would be those with the most extreme feminisation and would result in selection for separation of the sexes to give the familiar intercalary ear and terminal tassel.

The work by Mangelsdorf (1965) to simulate a primitive maize (defined in his terms) using existing stocks, independently of teosinte, suggested that evolution could be put into reverse and the results matched quite closely to the oldest and smallest cobs found in prehistoric cave deposits.

The converse conclusion therefore is that maize could have, but not necessarily did, evolve independently of teosinte.

The impact of finding Zea diploperennis

Positions about the evolution of maize with or without the involvement of teosinte were relatively entrenched when *Zea diploperennis* was discovered in Mexico in 1979. In that year Iltis *et al.* in the title of their article described it as a 'new teosinte', but in the next they spoke of 'this diploid wild maize'. This was somewhat surprising since the plant was a robust perennial lacking an ear, but foreshadowed a revised classification (Doebley and Iltis, 1980; Iltis and Doebley, 1980). This new classification of *Zea* took account of diversity in teosinte and emphasised its perceived closeness, in part, to maize.

The discovery of *Zea diploperennis* by Iltis *et al.* (1979) had another consequence. Mangelsdorf (1986) remarked

> The Wilkes hypothesis [that perennial teosinte had hybridised with an early domesticated maize to create various races of annual teosinte] appealed to me at once... It provided me with a powerful incentive to resume the experimental research I had retired from more than a decade earlier.

To test Wilkes' hypothesis, Mangelsdorf hybridised *Zea diploperennis* with an old Mexican popcorn variety of *Zea mays* 'Palomero Tuloqueno'. The results are set out in Fig. 12.4.

Mangelsdorf (1986) argued that the Wilkes hypothesis was confirmed and explained the derivation of both more robust maize types and annual teosinte. He added that annual teosinte, although not the parent of maize, was (by reason of frequent hybridisation with it) the purveyor of valuable genes to it originating in *perennial* teosinte.

There were to be yet more developments. Eubanks (1993) made reciprocal crosses between *Tripsacum dactyloides* and *Zea diploperennis*. Like its parents, the F_1 is perennial and cold tolerance has been contributed by the former. Since each parent has a functioning sessile and non-functioning pedicellate spikelet in the female part of the inflorescence, the occurrence of four rows of paired kernels was unexpected and was compared by Eubanks to the oldest archaeological maize specimens from Tehuacán, Mexico. Among F_2 segregates were plants with four or eight-rowed ears and others with single-rowed spikes.

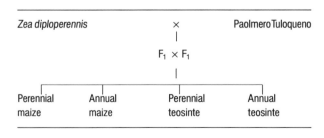

Figure 12.4. Perennial teosinte hybridised with maize (modified from Mangelsdorf, 1986).

Since the hybrid material is cross-fertile with maize it offers a means of introducing to it genes from *Tripsacum* including resistance to *Diabrotica virgifera*, root cutworm.

The enigma of heterosis

Hybrid vigour had long been recognised among plant and animal hybridists but it was Darwin (1876) who made of it a major systematic study. His book *The Effects of Self and Cross Fertilisation in the Vegetable Kingdom* was reviewed by the American, W. J. Beale and led eventually to the development of the hybrid corn industry. Despite the ability to manipulate hybrid vigour and the huge commercial enterprise it supports it is, fundamentally, a mystery. One can identify many individual hybrid effects that either do not persist or decline substantially from the F_2 onwards but their cause is unexplained.

Mangelsdorf (1965) attempted to identify a root cause of heterosis attributing it to the effects of ingress of germ plasm from *Tripsacum*. In his later work (Mangelsdorf, 1986) *Zea diploperennis* was assumed to have a central role in maize evolution and presumably (though not explicitly) *Tripsacum* might, as regards heterosis, have a diminished or non existent role. Heterosis itself remains an enigma.

Maize, the agricultural 'parvenu'

In no disparaging sense, maize is something of a parvenu, arising relatively recently from obscure and seemingly unpropitious origins to a position of immense significance. Indeed two things, particularly, are striking – namely the speed and the extent of change. The time interval is perhaps 5000 years and in that time a terminally produced ear weighing, fully grained, about 2.5 g has been transformed to a multiple-grain rowed, large naked-grained, fully husked intercalary ear weighing over 300 g. Against this background, both the genetic and the physiological instability described earlier become understandable to some extent. The impression is of a domesticated, though hardly tamed, C_4 grass, immensely active photosynthetically, driving the evolution of an ear which despite enlargement over numerous generations was never quite big enough for the job. (Or was it that an ear, capable of yet more enlargement, drew out of the leaves almost everything they could photosynthesise?)

Using, for a moment, the approach taken by Mangelsdorf (1986) it is possible to offer a linguistic analogy. Suppose the earliest *Zea mays* in cultivation is equated with Old English and perennial teosinte with Norman French, each largely separate until the Norman Conquest in 1066. Not only did Old English absorb its French counterpart, it retained a flexibility and an adaptability to become a world language. Maize, equally, having absorbed what teosinte had to offer has gone on to become a world crop. And just as English has its regional variations, so an American abroad would perhaps note that the maize there was not quite the same as that back home.

Maize Origin and Spread: An Oblique View from Ceramics

Early in this chapter mention was made of maize as a source of artistic inspiration. To a botanist in search of maize origins, of what significance is artistic or cultural tradition? One important consideration is that while much of the material recovered from plant debris is cob tissue, the whole ear or the grains it contained were of more interest to the artist and provided such representations were accurate they do, in fact, have something to offer the botanist. Of the many that might be used, two contrasted examples are presented here.

1.   San Pablo corn
This example is from Ecuador and involves corn about 4000 years ago. One item concerns a grain of corn that was included with the wet clay and was fully carbonised when the pot was fired. The grain, which had been in the early processes of germination, is preserved in a way that reveals detail of embryo, scutellum, endosperm and other features. Experimentally, it was found possible to re-create the artefact (Zevallos et al., 1977).

At another site in Ecuador, grains of maize were pressed either sideways or flat face as a pattern in to the rim of a newly made pot. Subsequent firing preserved the impression of the grain but the oil evaporated while other materials carbonised to leave a deposit of charcoal. Such impressions suggest strong similarity to an eight-rowed flint corn, Keallo ecuatoriano and found growing in Ecuador today (Zevallos et al., ibid.).

Other pots had impressions made using smaller grains. Again their carbonised remains were matched to several modern varieties, Pojoso chico, Chococero and Cariguil. Impressions made with both large and small-grained types suggest the long-term existence in Ecuador of these contemporary maize varieties or others very closely akin to them.

2.   The use of corn moulds
By fortunate circumstance it is the full ear that prompted so much Amerindian art. Moreover, if an actual ear were itself the object around which a two part mould were made, all manner of fine detail could be preserved and then imposed on the subsequent artefact. Such an approach is known for pre-Colombian cultures on the north coast of Peru and in the valley of Oaxaca in southern Mexico (Eubanks Dunn, 1984).

The moulds could be used to decorate both funerary urns and pots used every day. In Peru, Eubanks Dunn (*ibid.*) identified from ceramics some 19 races of maize on 35 jars. The jars were made over 1000 years ago but the maize races they represent or their close kin are, in most cases, growing today though not necessarily in Peru. When such materials are placed in cultural context they can provide evidence of plant exchange and social interaction.

A part of the interest in ceramics is thus the light shed on old varieties or 'ancient indigenous races', which survive both due to social preference and ecological adaptation. Bringing evidence of such races together with those from plant debris and ceramics contributes to the story of a maize ear steadily enlarging and with increased grain rows over the period of its domestication.

Agronomic Significance of the Enlarged Ear

In terms of human energy expended per ear during harvest, large ears rather than small ones would be preferred. This is even more so if yield per unit area can be increased rather than distributing the same yield over fewer ears.

The Future of Maize

Evans (1993) drew attention to the fact that ceilings to yield improvement had had, in the light of experience, to be revised repeatedly for the major cereals. Borrowing from Arthur C. Clarke, the science fiction writer, he suggested that in plant breeding, as in other fields of human endeavour, the impediments to progress are due either to failure of 'nerve' (courage) or failure of imagination. On this basis we need not attempt to define some ultimate maximum yield possible and indeed the clear implication is that it would be wiser not to. Maize yields will continue to rise and perhaps to an extent that surprises us, despite past increases.

At this point therefore, we leave maize to its uncertain but very promising future.

... And the Remainder of the Poaceae?

This concludes one individual's assessment of the grass family. It is to be hoped that the reader will have sensed something of its adaptedness, diversity and interest, but also, its malleability and usefulness. Perhaps too, contrary to his or her initial inclination, the reader will recognise a beckoning enterprise, in which they, without delay, now seek to participate.

References

For details of the man in the ice referred to in the Preface see Spindler (1994).

Anon. (1893) (Quoted by Collins, G.N. (1919)) *Meehans Monthly* 3, 105.

Anon. (1984) *Forage and Browse Plants for Arid and Semi-arid Africa.* IBPGR/RBG, Kew, pp. 293.

Anon. (1990) *Saline Agriculture. Salt-tolerant Plants for Developing Countries.* Rept. Nat. Res. Council. Nat. Academy Press, Washington DC, pp. 143.

Anon. (1993) *Vetiver Grass – A Thin Green Line Against Erosion.* Board on Science and Technology for International Development, National Research Council. Nat. Academy Press, Washington DC, pp. 171.

Appels, R. (1983) Chromosome structure in cereals: the analysis of regions containing repeated sequence DNA and its application to the detection of alien chromosomes introduced into wheat. pp. 229–255 in Kosuga, T., Meredith, C.P. and Hollaender, A. (Eds.) *Genetic Engineering in Plants.* Plenum Press, NY, pp. 499.

Arber, A. (1918) The phyllode theory of the monocotyledonous leaf with special reference to anatomical evidence. *Ann. Bot.* 32, 465–501.

Arber, A. (1934) *The Gramineae: a Study of Cereal, Bamboo and Grass.* Cambridge University Press, pp. 480.

Arencibia, A., Molina, P.R., de la Riva, G. and Selman-Housein, G. (1994) Production of transgenic sugar cane (*Saccharum officinarum* L.) plants by intact cell electroporation. *Plant Cell Repts.* 14, 305–309.

Asano, Y. and Ugaki, M. (1994) Transgenic plants of *Agrostis alba* obtained by electroporation-mediated direct gene transfer into protoplasts. *Plant Cell Repts.* 13, 243–246.

Avdulov, N.P. (1931) Karyo-systematische Untersuchung der Familie Gramineen. *Bull. Appl. Bot. Pl. Breed. Supp.* 44.

Bacon, C.W., Porter, J.K. and Robbins, J.D. (1986) Ergot toxicity from endophyte infected weed grasses: a review. *Agron. J.* 78, 106–116.

Baker, F.W.G. and Terry, J.P. (Eds.) (1991) *Tropical Grassy Weeds.* CAB International, Wallingford, pp. 203.

Barcelo, P. (1994) Transgenic cereal (*Tritordeum*) plants obtained at high efficiency by microprojectile bombardment of inflorescence tissue. *Plant Journal* 5, 583–592.

Barnes, P.W., Maggard, S., Holman, S.R. and Vergara, B.S. (1993) Intraspecific variation in sensitivity to UVB radiation in rice. *Crop Sci.* 33, 1041–1046.

Bashaw, E.C. and Hignight, K.W. (1990) Gene transfer in apomictic buffel grass through fertilisation of an unreduced egg. *Crop Sci.* 30, 571–575.

Beadle, G.W. (1980) The ancestry of corn. *Sci. Amer.* 242, 96–103.

Beaudry, J.R. (1951) Seed development following the mating *Elymus virginicus* L. × *Agropyron repens* L. Beauv. *Genetics* 36, 109–133.

Bedbrook, J.R., Jones, J., O'Dell, M., Thompson, R.D. and Flavell, R.B. (1980) A molecular description of telomeric heterochromatin in *Secale* species. *Cell* 19, 545–560.

Beetle, A.A. (1958) *Piptochaetium* and *Phalaris* in the fossil record. *Bull. Torr. Bot. Club* 85, 179–181.

Begg, J.E. and Wright, M.J. (1962) Growth and development of leaves from intercalary meristems in *Phalaris arundinacea* L. *Nature (London)* 194, 1097–1098.

Bentham, G. and Hooker, J.D. (1883) *Genera Plantarum* Vol. III, pp. 1074–1215.

Bessey, C. (1917) *The Phylogeny of Grasses*. Mich. Academic Sci. Rept. 19.

Bews, J.W. (1929) *The World's Grasses. Their Differentiation, Distribution, Economics and Ecology*. Longmans Green and Co., pp. 408.

Björk, S. (1967) Ecological investigations of *Phragmites communis*. *Folia Limnol. Scand.* 14.

Blaser, H.W. (1944) Studies in the morphology of Cyperaceae II. The Prophyll. *Amer. J. Bot.* 31, 53–64.

Boelens, M.H. (1994) Sensory and chemical evaluations of tropical grass oils. *Perfumer and Flavorist* 19, 29–48.

Boesel, R. and Schilcher, J. (1989) Composition of the essential oil of *Agropyron repens* rhizome. *Planta Medica* 55, 399–400.

Bogdan, A.V. (1961) Grass pollination by bees in Kenya. *Proc. Linn. Soc. London* 173, 57–60.

Bommineni, V.R. (1990) Maturation of stamens and ovaries on cultured ear inflorescences of maize (*Zea mays*). *Plant Tissue Organ Cult.* 23, 59–66.

Bommineni, V.R. and Greyson, R.I. (1987) In vitro culture of ear shoots of *Zea mays* and the effect of kinetin on sex expression. *Amer. J. Bot.* 74, 883–890.

Bommineni, V.R. and Greyson, R.I. (1990a) Regulation of flower development in cultured ears of maize (*Zea mays* L.). *Sex Plant Reprod.* 3, 103–115.

Bommineni, V.R. and Greyson, R.I. (1990b) Effects of gibberellic acid and indole 3 acetic acid on growth and differentiation of flowers in cultured ear inflorescences of maize (*Zea mays* L.). *Plant Sci.* 68, 239–247.

Bor, N.L. (1960) *The Grasses of Burma, Ceylon, India and Pakistan*. Pergamon, pp. 767.

Bosch, O.J.H. and Theunissen, J.D. (1992) Differences in the response of species on the degradation gradient in the semi-arid grasslands of southern Africa and the role of ecotypic variation. pp. 95–110 in Chapman, E.G. (Ed.) *Desertified Grasslands: Their Biology and Management*. Academic Press, pp. 360.

Boyd, L. (1931) Evolution in the monocotyledonous seedling: A new interpretation of the morphology of the grass embryo. *Trans. Bot. Soc. Edin.* 30, 286–303.

Braidwood, R.J. (1960) The Agricultural Revolution. *Sci. Amer.* 203, 130–148.

Bremer, G. (1961) Problems in breeding and cytology of sugar cane. IV. The origin of increase of chromosome number in species hybrids of *Saccharum*. *Euphytica* 10, 325–342.

Brink, R.A. and Cooper, D.C. (1944) The antipodals in relation to abnormal endosperm behviour in *Hordeum jubatum* × *Secale* cereal hybrid seeds. *Genetics* 29, 391–406.

Broome, S.W., Seneca, E.D. and Woodhouse, W.W. Jnr. (1986) Long term growth and development of transplants of the salt marsh grass *Spartina alterniflora Estuaries* 9, 63–74.

Brown, L.R. and Flavin, C. (1988) The earth's vital signs. pp. 1–21 in Starke, L. (Ed.) *The State of the World*. Penguin Books, Canada, pp. 237.

Brown, W.V. (1959) Grass leaf anatomy: its use in systematics. *Rec. Adv. in Bot. Montreal* 1, 105–108.

Brown, W.V. (1960) The morphology of the grass embryo. *Phytomorphology* 10, 215–223.

Bunting, A.H. (1971) Productivity and profit, or is your vegetative phase really necessary. *Ann. Appl. Biol.* 67, 265–272.

Burns, W. (1946) Corm and bulb formation in plants with special reference to the Gramineae. *Trans. Bot. Soc. Edin.* 34, 316–347.

Busri, N., Chapman, G.P. and Greenham, J. (1993) The embellum: a newly defined structure in the grass ovule. *Sex Plant Reprod.* 6, 191–198.

Calderon, C.E. and Söderstrom, T.R. (1973) Morphological and anatomical considerations of the grass subfamily Bambusoideae based on the new genus *Maclurolyra*. *Smithsonian Contributions to Botany* 11, 1–51.

Campbell, C.S. (1985) The subfamilies and tribes of the Gramineae (Poaceae) in the southeastern United States. *J. Arnold Arboretum* 66, 123–199.

Cass, D.D. and Jensen, W.A. (1970) Fertilisation in barley. *Amer. J. Bot.* 57, 62–70.

Cavalcante, V.A. and Dobereiner, J. (1988) A new acid-tolerant nitrogen-fixing bacterium associated with sugar cane. *Plant and Soil* 108, 23–31.

Celakovsky, L.J. (1897) Uber homologien der grasembryos. *Bot. Zig. (Leipzig)* 55, 141–174.

Cerling, T.E. and Quade, J. (1991) Fossil soils, grasses and carbon isotypes from Fort Ternan, Kenya: grassland or woodland? *J. Hum. Evol.* 21, 295–306.

Cerling, T.E., Wang, Y. and Quade, J. (1993) Expansion of C_4 ecosystems as an indicator of global ecological change in the late Miocene. *Nature (London)* 361, 344–345.

Chaloner, W.G. (1984) Plants, animals and time. *The Palaeobotanist* 32, 197–202.

Chandler, M.E.J. (1964) *Lower Tertiary Floras of Southern England IV.* A summary and survey of findings in the light of recent botanical observations. Brit. Mus. (Nat. His.) London, pp. 151.

Chao, C.S. (1989) *A Guide to Bamboos Grown in Britain*. Royal Botanic Gardens, Kew, pp. 47.

Chaplick, G.P. and Clay, K. (1988) Acquired chemical defences in grasses: the role of fungal endophytes. *OIKOS* 52, 309–318.

Chapman, G.P. (Ed.) (1990) *Reproductive Versatility in the Grasses*. Cambridge University Press, pp. 296.

Chapman, G.P. (Ed.) (1992a) *Desertified Grasslands: Their Biology and Management*. Academic Press, London, pp. 360.

Chapman, G.P. (Ed.) (1992b) *Grass Evolution and Domestication*. Cambridge University Press, pp. 390.

Chapman, G.P. (1992c) Apomixis and evolution. pp. 138–155 in Chapman, G.P. (Ed.) *Grass Evolution and Domestication*. Cambridge University Press, pp. 390.

Chapman, G.P. (1995) Grass inflorescence and spikelet culture: an appraisal. *Euphytica* 81, 121–129.

Chapman, G.P. and Busri, N. (1994) Apomixis in *Pennisetum:* an ultrastructural study. *Int. J. Pl. Sci.* 155, 492–497.

Chapman, G.P. and Peat, W.E. (1992) *An Introduction to the Grasses (Including Bamboos and Cereals)*. CAB International, Wallingford, pp. 111.

Chikkannaiah, P.S. and Mahlingappa, M.S. (1975) Antipodal cells in some members of Gramineae. *Curr. Sci.* 44, 22–23.

Clark, L.G. (1990) Diversity and biogeography of neotropical bamboos (Poaceae: Bambusoideae). *Acta Bot. Bras.* 4, 125–132.

Clark, L.G. and Fisher, J.B. (1987) Vegetative morphology of grasses: shoots and roots. pp. 37–45 in Söderstrom, T.R., Hilu, K.W., Campbell, C.S. and Barkworth, M.E. (Eds.) *Grass Systematics and Evolution*. Smithsonian Institute Press, Washington DC, pp. 472.

Clay, K. (1984) The effect of the fungus *Atkinsonella hypoxylon* (Clavicipitaceae) on the reproductive system and demography of the grass *Danthonia spicata. New Phytol.* 98, 165–175.

Clayton, W.D. (1975) Chorology of the genera of Gramineae. *Kew Bull.* 30, 111–132.

Clayton, W.D. (1981) Evolution and distribution of grasses. *Ann. Missouri Bot. Gdn.* 68, 5–14.

Clayton, W.D. (1990) The spikelet. pp. 32–51 in Chapman, G.P. (Ed.) *Reproductive Versatility in the Grasses*. Cambridge University Press, pp. 296.

Clayton, W.D. and Cope, T.A. (1980a) The chorology of old world species of Gramineae. *Kew Bull.* 35, 135–171.

Clayton, W.D. and Cope, T.A. (1980b) The chorology of North American species of Gramineae. *Kew Bull.* 35, 567–576.

Clayton, W.D. and Renvoize, S.A. (1986) *Genera Graminum: Grasses of the World. Kew Bull. Add. Ser.* 13. Royal Botanic Gardens Kew.

Clayton, W.D. and Renvoize, S.A. (1992) A system of classification for the grasses. pp. 338–353 in Chapman, G.P. (Ed.) *Grass Evolution and Domestication*. Cambridge University Press, pp. 390.

Clifford, H.T. (1961) Floral evolution in the family Gramineae. *Evolution* 15, 455–476.

Cocking, E.C., Davey, M.R. and Kothari, S.L. (1994) Nitrogen fixing cereals. *Landmark* (5) 3, Sept./Oct.

Collins, G.N. (1919) Structure of the maize ear as indicated in *Zea–Euchlaena* hybrids. *J. Agric. Res.* 17, 127–135.

Cooper, M. and Morris, D.W. (1983) *Grass Farming* 5th Edn. Farming Press, Ipswich, pp. 257.

Cope, T.A. and Simon, B.K. (1995) The chorology of Australasian grasses. *Kew Bull.* 50, 367–378.

Copeland, L.O. and Hardin, E.C. (1970) Outcrossing in rye grass *Lolium* spp. as determined by fluorescence tests. *Crop Sci.* 10, 254–257.

Cranwell, L.M. (1959) Fossil pollen from Seymour Island, Antarctica. *Nature (London)* 184, 1782–1785.

Creighton, H.B. and McClintock, B. (1931) A correlation of cytological and genetical crossing over in *Zea mays*. *Proc. Natl. Academy Sci. USA* 17, 492–497.

Crepet, W.L. and Feldman, G.F.D. (1991) The earliest remains of grasses in the fossil record. *Amer. J. Bot.* 78, 1010–1014.

Crowder, L.V. and Chheda, J.R. (1982) *Tropical Grassland Husbandry*. Longman, London, pp. 562.

Dahlgren, R.M.T., Clifford, H.T. and Yeo, P.F.G. (1985) *The Families of the Monocotyledons: Structure, Evolution and Taxonomy*. Springer-Verlag, Berlin, pp. 520.

Dhaliwal, H.S. and King, P.J. (1978) Direct pollination of *Zea mays* ovules *in vitro* with *Zea mays*, *Zea mexicana* and *Sorghum bicolor* pollen. *Theor. Appl. Genet.* 53, 43–46.

Da-lang Shen, Zi-fen Wang and Wu, M. (1987) Gene mapping on maize pachytene chromosomes by *in situ* hybridisation. *Chromosoma (Berl.)* 95, 311–314.

Darlington, C.D. (1932) *Recent Advances in Cytology*. Churchill, London, pp. 558.

Darlington, C.D. (1937) *Recent Advances in Cytology*. 2nd Edn. Churchill, London, pp. 671.

Darlington, C.D. (1956) *Chromosome Botany*. Allen and Unwin, London, pp. 186.

Darlington, C.D. (1969) *The Evolution of Man and Society*. Allen and Unwin, London, pp. 753.

Darlington, C.D. and Janaki-Ammal, E.K. (1945) *Chromosome Atlas of Cultivated Plants*. Allen and Unwin, London, pp. 397.

Darlington, C.D. and Mather, K. (1949) *The Elements of Genetics*. George, Allen and Unwin, London, pp. 446.

Darlington, C.D. and Wylie, A. (1955) *Chromosome Atlas of Cultivated Plants*. Allen and Unwin, pp. 520.

Darwin, C. (1859) *Origin of Species* 1st Edn. John Murray, London, pp. 671.

Darwin, C. (1868) *Animals and Plants Under Domestication*. 1st Edn. 2 vols. John Murray, London, pp. 411, 486.

Darwin, C. (1876) *The Effects of Self and Cross Fertilisation in the Vegetable Kingdom*. 1st Edn. John Murray, London, pp. 482.

Datta, S.K., Peterhaus, A., Datta, K. and Potrykus, I. (1990) Genetically engineered fertile indica rice recovered from protoplasts. *Biotech.* 8, 736–740.

Dawson, B., Peat, W.E. and Legg, C. (1995) *Grass ID. CLUES* Courseware, University of Aberdeen.

Dellaporta, S.L. and Calderon-Urrea, A. (1994) The sex determination process in maize. *Science* 266, 1501–1505.

Devos, K. and Gale, M. (1993) The genetic maps of wheat and their potential in plant breeding. *Outlook on Agriculture* 22, 93–99.

Dewald, C.L., Burson, B.L., de Wet, J.M.J. and Harlan, J.R. (1987) Morphology, inheritance and evolutionary significance of seed reversal in *Tripsacum dactyloides* (Poaceae). *Amer. J. Bot.* 74, 1055–1059.

de Wet, J.M.J. and Harlan, J.R. (1976) Systematics of Gramineae. *Phytologia* 33, 203–227.

de Wet, J.M.J., Harlan, J.R., Engle, L.M. and Grant, C.A. (1973) Breeding behaviour of maize *Tripsacum* hybrids. *Crop Sci.* 13, 254–256.

Dobereiner, J. (1961) Nitrogen fixing bacteria of the genus *Beijerinckia* Derx. in the rhizosphere of sugar cane. *Plant and Soil* 14, 211–217.

Dobereiner, J. and Day, J.H. (1975) Nitrogen fixation in the rhizosphere of tropical grasses. pp. 39–56 in Stewart, W.S.P. (Ed.) *Nitrogen Fixation by Free Living Organisms.* Cambridge University Press, pp. 471.

Dobereiner, J., Reis, V.M., Paula, M.A. and Olivares, F. (1992) Endophytic diazotrophs in sugar cane, cereals and tuber plants. pp. 671–676 in Palacios, R., Mora, J. and Newton, W.E. (Eds.) *New Horizons in Nitrogen Fixation.* Klewer Academic Pub.

Doebley, J.F. and Iltis, H. (1980) Taxonomy of *Zea* (Graminease) I. A subgeneric classification with key to taxa. *Amer. J. Bot.* 67, 982–993.

Doggett, J. (1988) *Sorghum* 2nd Edn. Longman, pp. 512.

Donald, C.M. (1968) The breeding of crop ideotypes. *Euphytica* 17, 388–403.

Dujardin, M. and Hanna, W.W. (1988) Production of 27, 28 and 26 (chromosome apomict hybrid derivatives between pearl millet ($2n = 14$) and *Pennisetum squamulatum* ($2n = 54$). *Euphytica* 38, 229–235.

Edwards, G.E. and Huber, S.C. (1981) The C_4 pathway. pp. 238–281 in Hatch, M.D. and Boardman, I.K. (Eds.) *The Biochemistry of Plants – a Comprehensive Treatise.* Vol. 8. Academic Press, NY, pp. 251.

Edwards, J.A. and Lewis Smith, R.I. (1988) Photosynthesis and respiration of *Colobanthus quitensis* and *Deschampsia antarctica* from the maritime Antarctic. *Brit. Antarctic Survey Bull.* B81B, 43–63.

Ehleringer, J.R., Sage, R.F., Flanagan, L.B. and Pearcy, R.W. (1991) Climatic change and the evolution of C_4 photosynthesis. *Trends in Ecol. and Evol.* 6, 95–99.

Eigsti, O.J. and Dustin, P. Jnr. (1955) *Colchicine in Agriculture, Medicine, Biology and Chemistry.* Iowa State College Press, Ames., Iowa, USA, pp. 470.

Elias, M.K. (1932) Grasses and other plants from the Tertiary rocks of Kansas and Colorado. *University of Kansas Sci. Bull.* 20, 333–367.

Elias, M.K. (1934) Zones of fossil herbs in the late Tertiary of the High Plains. *Geol. Soc. Amer. Proc.* 332 abst.

Elias, M.K. (1935) Tertiary grasses and other prairie vegetation from the High Plains of North America. *Amer. J. Sci.* 29, 24–33.

Elias, M.K. (1941) Late Tertiary prairie vegetation in Nebraska. pp. 19–20 and fig. 2 opp. p. 6. in *Guide for a Field Conference on the Tertiary Prairie Vegetation in Nebraska. Nebr. State Mus. Special Pub.* 2.

Elias, M.K. (1942) Tertiary prairie grasses and other herbs from the High Plains. *Geol. Soc. Amer. Special Paper (Regular Studies)* 41, 1–176.

Ellis, R.H., Hong, J.D. and Roberts, E.H. (1983) Procedure for the safe removal of dormancy from rice seed. *Seed Sci. and Tech. II,* 77–112.

Ellis, R.H., Hong, T.S. and Roberts, E.H. (1986) Quantal response of seed germination in *Brachiaria humidicola, Echinochloa turnerana, Eragrostis tef* and *Panicum maximum* to photon dose for the low energy reaction and the high irradiance reaction. *J. Exp. Bot.* 37, 742–753.

Ellis, R.P. (1981) Relevance of comparative leaf blade anatomy in taxonomic and functional research on the South African Poaceae. PhD thesis, Univ. of Pretoria, Pretoria (quoted by Hattersley and Watson (1992)).

Engell, K. (1994) Embryology of barley, IV. Ultrastructure of the antipodal cells of

Hordeum vulgare L. cv. 'Bomi' before and after fertilisation of the egg cell. *Sex Plant Reprod.* 7, 333–346.

Epis, R.C. and Chapin, C.E. (1975) Geomorphic and tectonic implications of the post-Laramide, late Eocene erosion surface in the southern Rocky Mountains. pp. 45–74 in Curtis, B.F. (Ed.) *Ceonozoic History of the Southern Rocky Mountains.* Geol. Soc. Amer., Boulder, Colorado.

Esau, K. (1964) *Plant Anatomy.* John Wiley and Sons, NY, pp. 735.

Eubanks, M. (1993) A cross between *Tripsacum dactyloides* and *Zea diploperennis. Maize Genetics Co-operative Newsletter* No 67, p. 39.

Eubanks Dunn, M. (1984) Ceramic evidence for the spread of prehistoric races of maize. *Res. Rept. Nat. Geog. Soc.* 16, 183–210.

Evans, L.T. (1993) *Crop Evolution, Adaptation and Yield.* Cambridge University Press, pp. 500.

Farooq, S. and Naqvi, S.H.M. (1987) Problems and prospects for rice × kallar grass hybridisation. *Pakistan J. Sci.* 30, 600–663.

Fisher, M.J., Rao, I.M., Ayaza, M.A., Lascano, C.E., Sanz, J.I., Thomas, R.J. and Vera, R.R. (1994) Carbon storage by introduced deep-rooted grasses in the South American savannas. *Nature (London)* 371, 236–238.

Frean, M.L., Barrett, D.R., Ariovich, D., Wolfson, M. and Cresswell, C.F. (1983) Intraspecific variability in *Alloteropsis semi-alata* (R. Br.) Hitchc. *Bothalia* 14, 901–913.

Freeling, M. (1984) Plant transposable elements and insertion sequences. *Ann. Rev. Pl. Physiol.* 35, 277–298.

Fuls, E.R. and Bosch, O.J.H. (1990) Environmental stress resistance and propagation studies of six *Cynodon dactylon* strains to assess reclamation suitability. *Landscape and Urban Planning* 19, 281–289.

Gaff, D.F. and Ellis, R.P. (1974) Southern African grasses with foliage that revives after dehydration. *Bothalia* 11, 305–308.

Gaff, D.F. and Latz, P.K. (1978) The occurrence of resurrection plants in the Australian flora. *Australian J. Bot.* 26, 485–492.

Galinat W.C. (1995) The origin of maize: grain of humanity. *Econ. Bot.* 49, 3–12.

Gallagher, E.J. (Ed.) (1984) *Cereal Production.* Butterworths, pp. 354.

Gibbs Russell, G.E. (1983) The taxonomic position of C$_3$ and C$_4$ *Alloteropsis semi-alata* (Poaceae) in southern Africa. *Bothalia* 14, 205–213.

Gill, B.S. and Kimber, G. (1974a) Giemsa C-Banding and the evolution of rye. *Proc. Natl. Academic Sci. USA* 71, 1247–1249.

Gill, B.S. and Kimber, G. (1974b) Giemsa C-Banding and the evolution of wheat. *Proc. Natl. Academy Sci. USA* 71, 4086–4090.

Good, R.O. (1964) *The Geography of Flowering Plants,* 3rd Edn. Longman, London, pp. 518.

Gopalakrishnan, C. and Nahan, M. (1977) Economic potential of bagasse as an alternative energy source: the Hawaiian experience. pp. 479–488 in Lockeretz, W. (Ed.) *Agriculture and Energy.* Academic Press, pp. 750.

Grassi, G. and Bridgewater, A. (1992) *Biomass for Energy and Environment: Agriculture and Industry in Europe. A Strategy for the Future.* European Union, Dir. Gen. for Sci., Res. and Dev., Brussels.

Greene, S.W. and Walton, D.W.H. (1975) Checklist for the sub Antarctic and Antarctic vascular flora. *Polar Rec.* 17, 473–484.

Greenham, J. and Chapman, G.P. (1990) Ovule structure and diversity. pp. 52–75 in Chapman, G.P. (Ed.) *Reproductive Versatility in the Grasses*. Cambridge University Press, pp. 296.

Griffin, G.F. (1992) Will it burn – should it burn? Management of the spinifex grassland of inland Australia. pp. 63–76 in Chapman, G.P. (Ed.) *Desertified Grasslands: Their Biology and Management*. Academic Press, pp. 360.

Grimshaw, R.G. (1992) The establishment of *Vetiveria zizanioides* in low rainfall areas. pp. 127–136 in Chapman, E.G. (Ed.) *Desertified Grasslands: Their Biology and Management*. Academic Press, pp. 360.

Grosser, D. and Liese, W. (1971) On the anatomy of Asian bamboos, with special reference to their vascular bundles. *Wood Science and Technology* 5, 290–312.

Grundbacher, F.J. (1963) The physiological function of the cereal awn. *Bot. Rev.* 29, 366–381.

Gupta, P.K. and Baum, B.R. (1989) Stable classification and nomenclature in the Triticeae: desirability, limitations and prospects. *Euphytica* 41, 191–197.

Gutterman, Y. (1992) Ecophysiology of Neger upland annual grasses. pp. 145–162 in Chapman, G.P. (Ed.) *Desertified Grasslands: Their Biology and Management*. Academic Press, pp. 360.

Hackel, E. (1890) Gramineae. pp. 1–96 in Engler and Prantl (Eds.) *Die Natürlichen Pflanzenfamilien*. Leipzig.

Hackel, E. (1892) *Monographia Festucarum Europearum*. Theodor Fischer, Kassel and Berlin.

Hackel, E. (1896) The true grasses. Translated from *Die Natütlicher Pflanzenfamilien* by Lawson-Scribner F. and Southworth, E.A. Archibald, Constance and Co. Westminster, pp. 228.

Hanna, W., Dujardin, M., Cozias-Akins, P., Lubbers, E. and Arthur, L. (1993) Reproduction, cytology and fertility of pearl millet × *Pennisetum squamulatum* BC_4 plants. *Jour. Hered.* 84, 213–216.

Harberd, D.J. (1962) Some observations of natural clones in *Festuca ovina*. *New Phytol.* 61, 85–100.

Harlan, J.R. (1989) Self perception and the origins of agriculture. pp. 5–23 in Swaminathan, M.A. and Koxhhar, S.L. (Eds.) *Plants and Society*. Macmillan.

Harlan, J.R. and de Wet, J.H.J. (1969) Sources of variation in *Cynodon dactylon* (L.) Pers. *Crop Sci.* 9, 774–778.

Harlan, J.R. and de Wet, J.H.J. (1971) Toward a rational classification of cultivated plants. *Taxon.* 20, 509–517.

Harlan, J.R., de Wet, J.H.J. and Richardson, W.L. (1969) Hybridisation studies with species of *Cynodon* from East Africa and Malagassy. *Amer. J. Bot.* 56, 944–950.

Harris, D.R. (1990) 3. Vavilov's concept of centres of origin of cultivated plants: its genesis and its influence on the study of agricultural origins. *Biol. J. Linn. Soc.* 39, 7–16.

Hartley, W. (1950) The global distribution of the tribes of the Gramineae in relation to historical and environmental factors. *Australian J. Agric. Res.* 1, 355–373.

Hartley, W. (1954) The agrostological index. *Australian J. Bot.* 2, 1–21.

Hartley, W. (1958a) Studies in the origin, evolution and distribution of the Gramineae I. The tribe Andropogoneae. *Australian J. Bot.* 6, 116–128.

Hartley, W. (1958b) II The tribe Paniceae. *Australian J. Bot.* 6, 343–357.

Hartley, W. (1961) IV The genus *Poa*. *Australian J. Bot.* 9, 152–161.

Hartley, W. (1973) V The subfamily Festucoideae. *Australian J. Bot.* 21, 201–234.

Hartley, W. and Slater, C. (1960) Evolution and distribution of the Eragrostoideae. *Australian J. Bot.* 8, 256–276.

Harz (1880–82) Bietrage zur Systematik der Gramineen. *Linnaea* XLIII, 1–30.

Hattersley, P.W. (1987) Variations in photosynthesis pathway. pp. 49–64 in Söderstrom, T.R., Hilv, K.W., Campbell, C.S. and Barkworth, M.E. (Eds.) *Grass Systematics and Evolution*. Smithsonian Institute Press, Washington DC, pp. 472.

Hattersley, P.W. and Stone, N.E. (1986) Photosynthetic enzyme activities in the C_3–C_4 intermediate *Neurachne minor* St Blake (Poaceae). *Australian J. Physiol.* 13, 399–405.

Hattersley, P.W. and Watson, L. (1992) Diversification of photosynthesis. pp. 58–116 in Chapman, G.P. (Ed.) *Grass Evolution and Domestication*. Cambridge University Press, pp. 390.

Hauber, D.P., White, D.A., Powers, S.P. and de Francesch, F.R. (1991) Isozyme variation and correspondence with unusual infra red reflectance patterns in *Phragmites australis* (Poaceae). *Plant Syst. Evol.* 178, 1–8.

Hawkes, J.G. (1983) *The Diversity of Crop Plants*. Harvard, pp. 184.

Hay, R.K.M. and Walker, A.J. (1989) *An Introduction to the Physiology of Crop Yield*. Longman, pp. 292.

Hayek, A. (1925) Zur Systematik der Gramineen. *Osterr. Bot. Zeitschr.* LXXIV (11–12), 249–255.

Hayman, D.L. (1992) The S–Z self-incompatibility system in the grasses. pp. 117–137 in Chapman, G.P. (Ed.) *Grass Evolution and Domestication*. Cambridge University Press, pp. 390.

He, J., Huang, L.-K., Chow, W.S., Whitecross, M.J. and Anderson, J.M. (1993) Effects of supplementary ultraviolet-B radiation on rice and pea plants. *Australian J. Pl. Physiol.* 20, 129–142.

He, J., Huang, L.-K., Chow, W.S., Whitecross, M.J. and Anderson, J.M. (1994) Responses of rice and pea plants to hardening with low doses of ultraviolet-B radiation. *Australian J. Pl. Physiol.* 21, 563–574.

Helbaeck, H. (1959) Domestication of food plants in the old world. *Science* 130, 365–372.

Helliwell, D.R., Buckley, G.P., Fordham, S.G. and Paul, T.A. (1996) Vegetation succession on a relocated woodland soil. *Foresty* (in press).

Herrera-Estrella, L. (1983) Chimeric genes as dominant selectable markers in plants cells. *EMBO. J.* 2, 987.

Heslop-Harrison, J. (1979) Pollen stigma interaction in the grasses: a brief review. *New Zealand J. Bot.* 17, 537–546.

Heslop-Harrison, J. (1987) Pollen germination and pollen tube growth. *Int. Rev. Cytol.* 10, 1–78.

Heslop-Harrison, Y. and Shivana, K.R. (1977) The receptive surface of the angiosperm stigma. *Ann. Bot.* 41, 1233–1258.

Hitch, P.A. and Sharman, B.C. (1971) The vascular pattern of festucoid grass axes, with particular reference to nodal plexi. *Bot. Gaz.* 132, 38–56.

Holttum, R.G. (1967) The bamboos of New Guinea. *Kew Bull.* 21, 263–292.

Hunter, R.B., Hunt, L.A. and Kannenberg, L.W. (1974) Photoperiod and temperature effects on corn. *Can. J. Pl. Sci.* 54, 71–78.

Hunziker, J.H., Wulff, A.F. and Söderstrom, T.R. (1989) Chromosome studies on *Anomochloa* and other Bambusoideae. *Darwiniana* 29, 41–45.

Huskins, C.L. (1931) The origin of *Spartina townsendii*. *Genetics* 12, 531–538.

Hussey, M.A., Bashaw, E.C., Hignight, K.W. and Dahmer, M.L. (1991) Influence of photoperiod on the frequency of sexual embryo sacs in facultative apomictic buffel grass. *Euphytica* 54, 141–145.

IIEDWRI (Internat. Ins. for Env. and Dev. World Resource Inst.) (1987) *World Resources*. NY.

Iltis, H. (1983) From teosinte to maize: the catastrophic sexual transmutation. *Science* 222, 886–894.

Iltis, H. and Doebley, J.F. (1980) Taxonomy of *Zea* (Gramineae) II. sub-specific categories in the *Zea mays* complex and a generic synopsis. *Amer. J. Bot.* 67, 994–1004.

Iltis, H., Doebley, J.F., Guzman, M.R. and Pazy, B. (1979) *Zea diploperennis:* a new teosinte from Mexico. *Science* 203, 186–188.

Islam ul Haq, M. and Khan, M.F.A. (1971) Reclamation of saline and alkaline soils by growing Kallar grass. *Nucleus* 8, 139–144.

Jackson, B.D. (1928) *A Glossary of Botanic Terms with their Derivation and Accent.* Duckworth, London, pp. 481.

Jacobs, S.W.L. (1992) Spinifex (*Triodia, Plectrachne, Symplectrodia* and *Monodia:* Poaceae) in Australia. pp. 47–62 in Chapman, G.P. (Ed.) *Desertified Grasslands: Their Biology and Management.* Academic Press, pp. 360.

Janzen, D.H. (1976) Why bamboos wait so long to flower. *Ann. Rev. Ecol. and Syst.* 7, 347–391.

Jones, D.F. (1934) Unisexual maize plants and their bearing on seed differentiation in other plants and animals. *Genetics* 19, 552–567.

Jones, M.B. and Lazenby, A. (1958) *The Grass Crop. The Physiological Basis of Production.* Chapman & Hall, London, pp. 369.

Jones, T.J. and Rost, T.L. (1989) Histochemistry and ultrastructure of rice (*Oryza sativa*) zygotic embryogenesis. *Amer. J. Bot.* 76, 504–520.

Judziewicz, E.J. and Söderstrom, T.R. (1989) Morphological, anatomical and taxonomic studies in *Anomochloa* and *Streptochaeta* (Poaceae: Bambusoideae). *Smithsonian Contribution to Bot.* I–III, 1–52.

Kellog, E.A. and Campbell, C.S. (1987) Phylogenetic analysis of the Gramineae. pp. 310–324 in Söderstrom, T.R., Hilu, K.W., Campbell, C.S. and Barkworth, M.E. (Eds.) *Grass Systematics and Evolution.* Smithsonian Institute Press, Washington DC, pp. 472.

Kenyon, K.M. (1978) *Archaeology in the Holy Land.* 4th Edn. Ernest Benn Ltd, London, pp. 360.

Kernick, M.D. (1990) An assessment of grass succession, utilisation and development in the arid zone. pp. 154–181 in Chapman, G.P. (Ed.) *Reproductive Versatility in the Grasses.* Cambridge University Press, pp. 296.

Kiniry, J.R., Ritchie, J.T. and Musser, R.L. (1983) Dynamic nature of the photoperiod response in maize. *Agron. J.* 75, 700–703.

Kleinhofs, A., Kilian, A., Saghai, M.A., Biyashev, R.M., Hayes, P., Chen, F.Q., Lapitan, N., Fenwich, A., Blake, T.K., Kanazin, V., Ananiev, E., Dahleen, L., Kudrna, D., Bollinger, J., Khapps, J., Lin, B., Sorrels, M., Heun, M., Franckowiek, J.D., Hoffman, D.S., Skadsden, R. and Steffensen, B.J. (1993) A molecular, isozyme and morphological map of the barley (*Hordeum vulgare*) genome. *Theor. Appl. Genet.* 86, 705–712.

Knox, R.B. and Singh, M.B. (1990) Reproduction and recognition phenomena in the Poaceae. pp. 220–239 in Chapman, G.P. (Ed.) *Reproductive Versatility in the Grasses.* Cambridge University Press, pp. 296.

Kranz, E., Bantor, J. and Lörz, H. (1991a) *In vitro* fertilisation of single isolated gametes of maize mediated by electrofusion. *Sex Plant Reprod.* 4, 12–16.

Kranz, E., Bantor, J. and Lörz, H. (1991b) Electrofusion-mediated transmission of cytoplasmic organelles through the *in vitro* fertilisation process, fusion by sperm cells with synergids and central cells and cell reconstitution in maize. *Sex Plant Reprod.* 4, 17–21.

Lagudah, E.S. and Appels, R. (1992) Wheat as a model system. pp. 225–265 in Chapman, G.P. (Ed.) *Grass Evolution and Domestication.* Cambridge University Press, pp. 390.

Laidler, K. and Laidler, L. (1992) *Pandas – Giants of the Bamboo Forests.* BBC Books, London, pp. 208.

Lamb, W. (1912) The phylogeny of grasses. *The Plant World* 15, 264–269.

Law, R., Bradshaw, A.D. and Putwain, P.D. (1977) Life history variation in *Poa annua. Evolution* 31, 233–246.

Lawler, D.W. (1993) *Photosynthesis: Molecular, Physiological and Environmental Processes.* 2nd Edn. Longman, pp. 318.

Lazarides, M. (1992) Resurrection grasses (Poaceae) in Australia. pp. 213–214 in Chapman, G.P. (Ed.) *Desertified Grasslands: Their Biology and Management.* Academic Press, pp. 360.

Lee, K.K., Wani, S.P., Yoneyama, T., Trimurtulu, N. and Harikrishnan, R. (1994) Associative N_2 fixation in pearl millet and sorghum: levels and response to inoculation. *Soil Sci. Plant Nutr.* 40, 477–484.

Linnaeus, C. (1753) *Species Plantarum.* Laurentii Salvii, Stockholm.

Linnaeus, C. (1767) *Genera Plantarum.* Thomae nob de Trattnern, Vienna.

Linnington, S., Bean, E.W. and Tyler, B.F. (1979) The effects of temperature upon seed germination in *Festuca pratensis* var. *apennina. J. Appl. Ecol.* 16, 933–938.

Lorenzo, J.L. and Gonzales, L. (1970) El mes antigio teosinte. *Bul de Instituto Nacional de Antro e Historia* No 42 Mex. City.

MacArthur, R.H. and Wilson, E.O. (1967) *The Theory of Island Biogeography.* Princeton University Press, pp. 203.

MacFadden, B. (1994) *Fossil Horses: Systematics, Palaeobiology and Evolution of the Family Equidae.* Cambridge University Press, pp. 384.

MacGinitie, H.D. (1953) Fossil plants of the Florissant beds, Colorado. *Publ. Carnegie Inst. Washington No* 599, 1–198.

MacNeish, R.S. (1985) The archaeological record on the problem of the domestication of corn. *Maydica* 30, 171–178.

Malkin, E. and Waisel, Y. (1986) Mass selection for salt resistance in Rhodes grass (*Chloris gayana*). *Physiologia Plantarum* 66, 443–446.

Mangelsdorf, P.C. (1958) Ancestor of corn. *Science* 128, 1313.

Mangelsdorf, P.C. (1961) Introgression in maize. *Euphytica* 10, 156–168.

Mangelsdorf, P.C. (1965) The evolution of maize. pp. 23–49 in Hutchinson, Sir J. (Ed.) *Essays on Crop Plant Evolution.* Cambridge University Press, pp. 204.

Mangelsdorf, P.C. (1986) The origin of corn. *Sci. Amer.* 255, 80–86.

Mangelsdorf, P.C., Macneish, R.S. and Galinat, W.C. (1964) Domestication of corn. *Science* 143, 538–545.

Mathauda, G.S. (1952) Flowering habits of the bamboo – *Dendrocalamus strictus. Ind. For.* 78, 86–88.

Mayer, C. (1937) *Soil Erosion and Land Utilisation in the Akamba Reserve (Machalos).* Report to the Department of Agriculture. Mss. Afr. 5755 Rhodes House Library, Oxford.

Mazzoleni, S. and Riccardi, M. (1993) Primary succession on the cone of Vesuvius. pp. 101–112 in Miles, J. and Walton, D.H.W. (Eds.) *Primary Succession on Land.* Spec. Pub. No. 12. Brit. Ecol. Soc. Blackwell, Oxford.

Mazzoleni, S., Riccardi, M. and Aprile, G.G. (1989) Aspetti pionieri della vegetazione del Vasuvio. *Ann Bot. (Roma)* 47, Suppl. 6, 97–110.

McClelland, T.B. (1928) Studies in the photoperiodism of some economic plants. *J. Agric. Res.* 37, 603–628.

McClintock, B. (1950) The origin and behaviour of mutable loci in maize. *Proc. Natl. Academy Sci. USA* 36, 344–355.

McClure, F.A. (1966) *The Bamboos: A Fresh Perspective.* Harvard University Press, pp. 347.

McClure, F.A. (Ed. Söderstrom, T.R.) (1973) Genera of bamboo native to the New World (Gramineae: Bambusoideae). *Smithsonian Contrib. to Bot.* 9, 6–8 and 83–87.

McClure, F.A. and Söderstrom, T.R. (1972) The agrostogical term anthecium. *Taxon* 21, 153–154.

McMullen, J.T., Morgan, R. and Murray, R.B. (1983) *Energy Resources.* 2nd Edn. Edwin Arnold, pp. 191.

McNeilly, T. and Antonovics, J.A. (1968) Evolution in closely adjacent populations. III *Agrostis tenuis* on a small copper mine. *Heredity* 23, 205–218.

Meyer, F.G. (1980) Carbonised food plants of Pompeii, Herculaneum and the villa at Torre Annunziata. *Econ. Bot.* 34, 401–437.

Meyer, S.E., Beckstead, J., Allen, P.S. and Pullman, H. (1995) Germination ecophysiology of *Leymus cinereus* (Poaceae). *Int. J. Plant Sci.* 156, 206–215.

Miller, T.E. (1987) Systematics and evolution. pp. 1–30 in Lupton, F.G.H. (Ed.) *Wheat Breeding: Its Scientific Basis.* Chapman & Hall, London, pp. 566.

Milthorpe, F.L. and Ivins, J.D. (1966) *The Growth of Cereals and Grasses.* Butterworths, London, pp. 359.

Mitchley, J. (1994) Grassland. pp. 79–95 in Watt, T.A. and Chapman, G.P. (Eds.) *The Natural History of a Country Estate: Wye College Kent.* Wye College Press, pp. 195.

Mitchley, J. and Buckley, G.P. (1994) Channel Tunnel Project. Vegetation establishment on Channel Tunnel spoil. Monitoring of the 1992 Permanent Sowing. Rept. to Contractor.

Mitchley, J. and Buckley, G.P. (1995) Habitat creation on the Channel Tunnel spoil in a maritime environment at Dover, UK. *Brit. Ecol. Soc. Conf. Leicester. March 1995,* pp. 1–6.

Mladá, J. (1974) The histological structure of the grass embryo and its significance for the taxonomy of the family Poaceae. *Acta Universitatis Carolinae – Biologica,* 51–156.

Mogensen, H.L. (1990) Fertilisation and early embryogenesis. pp. 76–99 in

Chapman, G.P. (Ed.) *Reproductive Versatility in the Grasses.* Cambridge University Press, pp. 296.

Mogensen, H.L. and Rusche, M. (1985) Quantitative ultrastructural analysis of barley sperm. I. Occurrence and mechanism of cytoplasm and organelle reduction and the question of sperm dimorphism. *Protoplasma* 128, 1–13.

Morley, R.J. and Richards, K. (1993) Gramineae cuticle: a key indicator of Late Cenozoic climatic change in the Niger Delta. *Rev. Palaeobot. and Palynol.* 77, 119–127.

Moss, F.I. and Heslop-Harrison, J. (1960) Photoperiod and pollen sterility in maize. *Ann. Bot.* 32, 833–846.

Muller, J. (1981) Fossil pollen records of extant angiosperms. *Bot. Rev.* 47, 1–145.

Murdoch, A.H. and Ellis, R.H. (1992) Longevity, viability and dormancy. pp. 193–229 in Fenner, M. (Ed.) *Seeds: The Ecology of Regeneration in Plant Communities.* CAB International, Wallingford, pp. 373.

Nadgauda, R.S., Parasharami, V.A. and Mascarenhas, A.F. (1990) Precocious flowering and seeding behaviour in tissue cultured bamboos. *Nature (London)* 334, 335–336.

Naylor, R.E.L. (1972) Biological flora of the British Isles. List of British vascular plants (1958) No 708.1. *Alopecurus myosuroides* Huds. (*A. agrostis* L.). *J. Ecol.* 60, 611–622.

Nehra, N.S., Chibbar, R.N., Leung, N., Caswell, K., Mallard, C., Steinhauer, L., Bagg, M. and Kartha, K.K. (1994) Self-fertile transgenic wheat plants regenerated from isolated scutellar tissues following micro-projectile bombardment with two distinct gene constructs. *The Plant Journal* 5, 285–297.

Nitsch, C., Mornan, K. and Godard, M. (1986) Intergeneric crosses between *Zea* and *Pennisetum* reciprocally by *in vitro* methods. pp. 53–58 in Mulcahy, D.L., Malcahy, G.B. and Cottaviana, E. (Eds.) *Biotechnology and Ecology of Pollen.* Springer-Verlag, NY, pp. 530.

Ozias-Atkins, P., Lubbers, E.L., Hanna, W.W. and McNay, J.W. (1993) Transmission of the apomictic mode of reproduction in *Pennisetum*; co-inheritance of the trait and molecular markers. *Theor. Appl. Genet.* 85, 632–638.

Page, J.S. (1978) A scanning electron microscope study of grass pollen. *Kew Bull.* 32, 313–319.

Palmer, P.G. (1976) Grass cuticles: a new palaeocological tool for East African lake sediments. *Can. J. Botany* 54, 1725–1734.

Pareddy, D.R. and Greyson, R.I. (1985) *In vitro* culture of immature tassels of an inbred field variety of *Zea mays* Oh : 43. *Plant Cell Tissue and Organ Cult.* 5, 119–128.

Pareddy, D.R., Greyson, R.I. and Walden, D.B. (1987) Fertilisation and seed production with pollen from *in vitro* cultured maize tassels. *Planta* 170, 141–143.

Parsons, J.J. (1970) The 'Africanisation' of the new world tropical grasslands. *Tubinger Geographische Studien* 34, 141–153.

Pearce, F. (1995) Dead in the water. *New Scientist* 145, 26–31.

Peart, M.H. (1979) Experiments on the biological significance of the morphology of seed dispersal units in grasses. *J. Ecol.* 67, 843–867.

Peart, M.H. (1981) Further experiments on the biological significance of the morphology of the seed dispersal units in grasses. *J. Ecol.* 69, 425–436.

Peart, M.H. (1984) The effects of morphology orientation and position of grass diaspores on seedling survival. *J. Ecol.* 72, 437–453.

Peart, M.H. and Clifford, H.T. (1987) The influence of diaspore morphology and soil surface properties on the distribution of grasses. *J. Ecol.* 75, 569–576.

Peters, N.C.B. (1982) The dormancy of wild oat seed (*Avena fatua* L.) from plants grown under various temperature and soil moisture conditions. *Weed Research* 22, 205–212.

Peyre de Fabrégues, B. (1992) Observations on the ebb and flow of native grass species in the area of the Ekrafane Ranch, Sahel. pp. 37–46 in Chapman, G.P. (Ed.) *Desertified Grasslands: Their Biology and Management.* Academic Press, pp. 360.

Philipson, M.N. (1978) Apomixis in *Cortaderia jubata* (Gramineae). *New Zealand J. Bot.* 16, 45–59.

Poinar, G.O. Jnr. (1992) *Life in Amber.* Stanford University Press, pp. 350.

Poinar, G.O. Jnr. and Columbus, J.T. (1992) Adhesive grass spikelet with mammalian hair in Dominican amber – first fossil evidence of epizoochory. *Experientia* 48, 906–909.

Polowick, P.L. and Greyson, R.I. (1982) Another development, meiosis and pollen formation in *Zea* tassels cultured in defined liquid medium. *Plant Sci. Lett.* 26, 139–145.

Polowick, P.L. and Greyson, R.I. (1984) The relative efficiency of cytokinins in the development of normal spikelets on cultured tassels of *Zea mays. Can. J. Bot.* 62, 830–834.

Polowick, P.L. and Greyson, R.I. (1985) Microsporogenesis and gametophyte maturation in cultured tassels of *Zea mays. Can. J. Bot.* 63, 2196–2199.

Porter, J.K., Canon, C.W., Cutler, H.G., Arrondale, R.F. and Robbins, J.D. (1985) *In vitro* auxin production by *Balansia. Phytochemistry* 24, 1429–1431.

Powell, R., Plattner, R., Yates, S., Clay, K. and Leuchtmann, A. (1990) Ergobalansine: a new ergot-type peptide alkaloid isolated from *Cenchrus echinatus* infected with *Balansia obtecta. Pl. Med.* 56, 518.

Prat, H. (1992) L'epiderme des Gramineés. Étude anatomique et systematique. *Ann. Sci. Nat. Bot. Ser.* 10, 14, 117–324.

Prat, H. (1960) Revue d'agrostologie: vers une classification naturelle des Gramineés. *Bull. Soc. Bot. Fr.* 107, 32–79.

Prendergast, H.D.V., Hattersley, P.W., Stone, N.E. and Lazarides, M. (1986) C_4 acid decarboxylation type in *Eragrostis* (Poaceae): patterns of variation in chloroplast position, ultrastructure and geographical distribution. *Plant Cell and Environment* 9, 333–344.

Prendergast, H.D.V., Hattersley, P.W. and Stone, N.E. (1987) New structural/biochemical associations in leaf blades of C_4 grasses (Poaceae). *Australian J. Pl. Physiol.* 14, 403–420.

Prendergast, H.D.V., Stone, N.E. and Tattersley, P.W. (1988) Leaf structure and C_4 acid decarboxylation enzymes in × *Cynochloris* spp. (Poaceae) inter-generic hybrids between species of different C_4 type. *Bot. J. Linn. Soc.* 96, 381–389.

Prendergast, H.D.V., Linington, S. and Smith, R.D. (1992) The Kew seed bank and the collection, storage and utilisation of arid and semi-arid zone grasses. pp. 235-250 in Chapman, G.P. (Ed.) *Desertified Grasslands: Their Biology and Management.* Academic Press, pp. 360.

Probert, R.J. (1992) The role of temperature in germination ecophysiology. pp. 285–325 in Fenner, M. (Ed.) *Seeds. The Ecology and Regeneration of Plant Communities.* CAB International, Wallingford.

Ragland, J.L., Hatfield, A.L. and Benoit, G.R. (1966) Photoperiod effects on the ear components of corn *Zea mays. Agron. Jour.* 58, 455–456.

Raloff, J. (1993) Corn's slow path to stardom. *Science News* 143, 248–251.

Rana, R.S. and Parkash, V. (1980) Kallar grass grows well on alkali soils. *Indian Farming* (July) 13, 17 and 19.

Raven, C.E. (1950) *John Ray, Naturalist, His Life and Works.* Cambridge University Press, pp. 506.

Ray, J. (1682) *Methodus Plantarum.* pp. 166 + index.

Ray, J. (1704) *The Wisdom of God Manifest in the Works of Creation.* 4th Edn., pp. 464.

Reeder, J.R. (1957) The embryo in grass systematics. *Amer. J. Bot.* 44, 757–768.

Reeder, J.R. (1962) The bambusoid embryo: a reappraisal. *Amer. J. Bot.* 49, 639–641.

Reid, D., Taylor, A., Jinchu, H. and Zisheng, Q. (1991) Environmental influences on bamboo *Bashiana fangiana* growth and implications for giant panda conservation. *J. App. Ecol.* 28, 855–868.

Ren Jizhou, Zhu Xingu and Yan Shunguo (1992) The ecological role of *Puccinellia chinampoensis* on saline soil in arid inland regions of China. pp. 137–144 in Chapman G.P. (Ed.) *Desertified Grasslands: Their Biology and Management.* Academic Press, pp. 360.

Renvoize, S.A. and Clayton, W.D. (1992) Classification and evolution of grasses. pp. 3–37 in Chapman, G.P. (Ed.) *Grass Evolution and Domestication.* Cambridge University Press, pp. 390.

Renvoize, S.A., Cope, T.A., Cook, G.E.M., Clayton, W.D. and Wickens, G.E. (1992) Distribution and utilisation of grasses in arid and semi-arid regions. pp. 3–16 and 323–332 (Appendix) in Chapman, G.P. (Ed.) *Desertified Grasslands: Their Biology and Management.* Academic Press, pp. 360.

Retallack, G.H., Dugas, D.P. and Bestland, E.A. (1990) Fossil soils and grasses of a Middle Miocene East African grassland. *Science* 247, 1325–1328.

Rhoades, M.M. (1945) On the genetic control of mutability in maize. *Proc. Natl. Academy Sci. USA* 31, 91–95.

Ricciardi, M., Aprile, G.G., LaValva, V. and Caputo, G. (1986) La Flora del Somma-Vesuvio. *Bolletino Società Naturalisti in Napoli Officine Grafiche Napoletane* 96, 3–121.

Richards, A.J. (1990) The implications of reproductive versatility for the structure of grass populations. pp. 131–153 in Chapman, G.P. (Ed.) *Reproductive Versatility in the Grasses.* Cambridge University Press, pp. 296.

Ridley, H.N. (1930) *The Dispersal of Plants throughout the World.* L. Reeve, Ashford, pp. 744.

Robinson, S.P. and Walker, D.A. (1981) Photosynthetic carbon reduction cycle. pp. 194–236 in Hatch, M.D. and Boardman, N.K. (Eds.) *The Biochemistry of Plants – A Comprehensive Treatise Vol. 8.* Academic Press, NY.

Romo, J.T. (1994) Wolf plant effects on water relations, growth and productivity in Crested Wheatgrass. *Can. J. Pl. Sci.* 74, 767–771.

Rost, T.L. and Lersten, N.R. (1973) A synopsis and selected bibliography of grass caryopsis, anatomy and fine structure. *Iowa State J. Res.* 48, 47–87.

Roth, J. (1959) Histogenesis and morphological interpretation of the grass embryo. *Recent Adv. Bot. Montreal* 1, 96–99.

Russell, S.D. (1986) Dimorphic sperm cells, cytoplasmic transmission and preferential fertilisation in *Plumbago zeylanica*. pp. 69–116 in Mantell, S.H., Chapman, G.P. and Street, P.F.S. (Eds.) *The Chrondriome: Chloroplast and Mitochondrial Genomes*. Longman, pp. 310.

Sandhu, G.R. and Malik, J.(1975) Plant succession – a key to the utilisation of saline soils. *Nucleus* 12, 35.

Sandhu, G.R., Aslam, S., Salim, M., Settar, A., Qureshi, R.H., Ahmad, N. and Wyn Jones, R.G. (1981) The effect of salinity on the yield of *Diplachne fusca* (Kallar Grass). *Plant, Cell and Environment* 4, 177–181.

Sargant, E. and Arber, A. (1915) The comparative morphology of the embryo and seeding in the Gramineae. *Ann. Bot. (London)* 89, 161–222.

Schaller, G.B. (1993) *The Last Panda*. University of Chicago Press, Chicago and London, pp. 291.

Schell, J.H.N., Kieft, H. and van Klammeren, A.A.M. (1984) Interactions between embryo and endosperm during early developmental stages of maize caryopses (*Zea mays*). *Can. J. Bot.* 62, 2842–2853.

Schellenberg, G. (1922) Die systematische Gliederung der Gramineen. *Bot. Archiv.* Bd1. Aeft 5.

Shotwell, J.A. (1961) Late Tertiary biogeography of horses in the northern Great Basin. *J. Paleontol.* 35, 203–217.

Simpson, G.G. (1951) *Horses*. Oxford University Press, pp. 247.

Simpson, G.M. (1990) *Seed Dormancy in Grasses*. Cambridge University Press, pp. 297.

Smart, M.G. and O'Brien, T.P. (1983) The development of the wheat embryo in relation to the neighbouring tissues. *Protoplasma* 114, 1–13.

Smith, B.D. (1994) *The Emergence of Agriculture*. Pub. Scientific American, pp. 231.

Smith, R.C., Prezelin, B.B., Baber, K.S., Bidigare, R.R., Boucher, N.P., Coley, T., Karentz, D., Macintyre, S., Matlick, H.A., Menzies, D., Gudrusek, M., Wan, Z. and Waters, K.J. (1992) Ozone depletion, ultraviolet radiation and phytoplankton biology in Antarctic waters. *Science* 255, 952–959.

Smith, R.I.L. and Prince, P.A. (1985) The natural history of Beauchênne island. *Biol. J. Linn. Soc. (London)* 24, 233–283.

Söderstrom, T.R. (1971) Insect pollination in tropical rain forest grasses. *Biotropica* 3, 1–16.

Söderstrom, T.R. (1981) The grass subfamily Centothecoideae. *Taxon* 30, 614–615.

Söderstrom, T.R. and Calderon, C.E. (1971) Insect pollination in tropical forest grasses. *Biotropica* 3, 1–16.

Söderstrom, T.R. and Calderon, C.E. (1980) In search of the primitive bamboos. *Nat. Geog. Soc. Res. Rept.* 12, 647–654.

Söderstrom, T.R. and Londoño, X. (1988) A morphological study of *Alvinia* (Poaceae: Bambuseae): a new Brazilian genus with fleshy fruits. *Amer. J. Bot.* 75, 819–839.

Söderstrom, T.R., Hilu, K.W., Campbell, C.S. and Barkworth, M.E. (Eds.) (1987) *Grass Systematics and Evolution*. Smithsonian Institute Press, Washington DC, pp. 472.

Speller, C.S. (1993) The potential for growing biomass crops for fuel on surplus land in the UK. *Outlook on Agriculture* 22, 23–29.

Spindler, K. (1994) *The Man in the Ice*. Translated by Ewald Cosers. Weidenfeld and Nicolson, London, pp. 305.

Stace, C.A. (1975) *Hybridisation and the Flora of the British Isles*. Academic Press, pp. 626.

Stace, C.A. (1991) *New Flora of the British Isles*. Cambridge University Press, pp. 1226.

Stapf, O. (1904) Anthoecium. *Botanischer Jarhbuch, Engler*. 35, 64–68.

Stapf, O. (1926) (Anthoecium). Exchange of corresp. with E.P. Phillips. Royal Botanic Gardens, Kew, Archives.

Stearn, W.T. (1957) The species concept of Linnaeus. pp. 151–161 in Vol. 1 Facsimile Edition. *Species Plantarum* Linnaeus C. 1753. Pub. Ray Society.

Stebbins, G.L. (1950) *Variation and Evolution in Plants*. Columbia University Press, pp. 643.

Stebbins, G.L. (1956) Taxonomy and the evolution of genera with special reference to the family Gramineae. *Evolution* 10, 235–245.

Stebbins, G.L. (1965) Colonising species of the native California flora. pp. 173–195 in Baker, H.G. and Stebbins, G.L. (Eds.) *The Genetics of Colonising Species*. Academic Press, London, pp. 588.

Stebbins, G.L. (1981) Co-evolution of grasses and herbivores. *Ann. Missouri Bot. Gard.* 68, 75–86.

Stebbins, G.L. (1987) Grass systematics and evolution: past present and future. pp. 359–367 in Söderstrom, T.R., Hilu, K.W., Campbell, C.G. and Barkworth, M.E. (Eds.) *Grass Systematics and Evolution*. Smithsonian Institute Press, Washington DC, pp. 472.

Strelkova, O. (1938) Polyploidy and geographo-systematic groups in the genus *Alopecurus*. *Cytologia* 8, 468–480.

Struik, P.C. (1982) Effect of a switch in photoperiod on reproductive development of temperate hybrids of maize. *Netherlands J. Agri. Sci.* 30, 69–83.

Symons, S., Simpson, G.M. and Adkins, S.W. (1987) Secondary dormancy in *Avena fatua*: effect of temperature and after ripening. *Physiol. Plantarum* 70, 419–426.

Taylor, A. (1988) Regeneration from seed of *Sinarundinaria fangiana*, a bamboo, in the Wolong giant panda reserve, Sichuan, China. *Amer. J. Bot.* 75, 1065–1073.

Taylor, A., Reid, D. and Zisheng, Q. (1991) Bamboo dieback: an opportunity to restore panda habitat. *Env. Cons.* 18, 166–168.

Taylorson, R.B. and Di Nola, L. (1989) Increased phytochrome responsiveness and a high temperature transition in barnyard grass. (*Echinochloa crus-galli*) seed dormancy. *Weed Sci.* 37, 335–338.

Tefera, H. (1992) *In vitro* development of t'ef (*Eragrostis tef*) seeds within detached spikelets. pp. 305–307 in Chapman, G.P. (Ed.) *Desertified Grasslands: Their Biology and Management*. Academic Press, pp. 360.

Teramura, A.H., Zinska, L.H. and Sztein, A.E. (1991) Changes in growth and photosynthetic capacity of rice with increase UV-B radiation. *Physiol. Plantarum* 83, 373–380.

Theophrastus ΠΕΡΙ ΦΥΤΩΝ ΙΣΤΟΡΙΑΕ – *Enquiry into Plants* translated by Sir Arthur Hort 1916.

Therdyothin, A., Bhattacharaya, S.C. and Chirarattananon, S. (1992) Electricity generation potential of Thai sugar mills. *Energy Sources* 14, 367–380.

Thomas, D.S.J. and Middleton, N.J. (1994) *Desertification – Exploding the Myth*. Wiley, Chichester, pp. 194.

Thomasson, J.R. (1980) Palaeoagrostology: a historical review. *Iowa State J. Res.* 54, 301–317.

Thomasson, J.R. (1987) Fossil grasses: 1820–1986 and beyond. pp. 159–167 in Söderstrom, T.R., Hilu, K.W., Campbell, C.G. and Barkworth, M.E. (Eds.) *Grass Systematics and Evolution*. Smithsonian Institute Press, Washington DC.

Thomasson, J.R., Nelson, M.E. and Zakrzewski, R.J. (1986) A fossil grass (Gramineae: Chloroidoideae) from the Miocene with Kranz anatomy. *Science* 233, 876–878.

Tiffen, M., Mortimere, M. and Gichuki, F. (1994) *More People, Less Erosion. Environmental Recovery in Kenya*. John Wiley and Sons, pp. 311.

Timberlake, L. (1985) *Africa in Crisis: the Causes, the Cures of Environmental Bankruptcy*. Tinker, J. (Ed.) IIED, pp. 232.

Tomlinson, P.B. (1970) Monocotyledons – toward an understanding of their morphology and anatomy. *Adv. in Bot. Res.* 3, 208–292.

Truswell, E.M. (1993) Vegetation changes in the Australian Tertiary in response to climatic and phytogeographic forcing factors. (Nancy Burbridge Memorial Lecture). *Australian Syst. Bot.* 6, 533–557.

Turpin, P.J.F. (1819) Mémoire sur l'inflorescence des Graminées et des Cyperées. comparée avec cell des autres végétaux sexiferes; suivi de quelques observations sur les disques. *Mem. Mus. His. Nat. Paris* 4, 67.

Umali, D.L. (1993) *Irrigation – Induced Salinity – A Growing Problem for Development and the Environment*. World Bank Tech. Paper No. 215, pp. 78.

van der Toorn, J. (1972) 48. *Variability of Phragmites australis (cav.) Trin. ex Steudel in Relation to the Environment*. Van See Tot. Land, pp. 122.

van Tieghem, P. (1897) Morphologie de l'embryon et de la plantule chez les Graminées et les Cyperacées. *Ann. Sci. Nat. Bot.* VIII, 259–309.

Vavilov, N.I. (1920) [The law of homologous series in heritable variation]. Translation by Star Chester K. (1949/50) in [*The Origin, Variation, Immunity and Breeding of Cultivated Plants*]. *Chronica Botanica* 13, 1–364.

Velenovsky, J. (1907) Všeobecná botanika Srovnavaci morphologie II Praha. (quoted by Mlada, J. (1974)).

Venkatash, C.S. (1984) Dichogamy and breeding system in a tropical bamboo *Ochlandra travancorica*. *Biotropica* 16, 309–312.

Vicari, M. and Bazely, D.R. (1993) Do grasses fight back? The case for antiherbivore defenses. *Trends in Ecol. and Evol.* 8, 137–141.

Voorhies, M.R. and Thomasson, J.R. (1979) Fossil grass anthecia with Miocene rhinoceros skeletons: diet of an extinct species. *Science* 206, 331–333.

Walker, D. (1992) *Energy Plants*. Oxygraphics, pp. 277.

Wallace, L.E., McNeal, F.H. and Berg, M.A. (1974) Resistance to both *Oulema* and *Cephus cinctus* in pubescent-leaved and solid-stemmed wheat selections. *J. Econ. Entom.* 67, 106–107.

Wan, Y. and Lemaux, P.G. (1994) Generation of large numbers of independently transformed fertile barley plants. *Plant Physiol.* 104, 37–48.

Wang, K. (1990) Ethnobotanical studies of bamboo resources in Mengsong, Xishuangbanna, Yunnan, China. *2nd. Int. Cong. Ethnobiology, Kunming, Yunnan, China*. pp. 1–7.

Wang, K., Xue, J.C.S., Pel, S.A.K. (1993) Ethnobotanical studies of bamboo resources in Xishuangbanna, Yunnan, China. *Papers of Tropical Botany Resources, Yunnan University Press* 8, 47–65.

Watanbe, K. (1955) Studies in the germination of grass pollen. 1. Liquid exudation of the pollen on the stigma before germination. *Bot. Mag. Tokyo* 68, 40–44.

Watson, L. (1990) The grass family Poaceae. pp. 1–31 and Appendix: World Grass Genera pp. 258–265 in Chapman G.P. (Ed.) *Reproductive Versatility in the Grasses.* Cambridge University Press, pp. 296.

Watson, L. and Dallwitz, M.J. (1988) *Grass Genera of the World: Interactive Identification and Information Retrieval.* Research School of Biological Sciences, Australian National University, Canberra, pp. 45.

Watson, L. and Dallwitz, M.J. (1992) *The Grass Genera of the World.* CAB International, Wallingford, pp. 1024.

Wayman, M. and Parekh, S.R. (1990) *Biotechnology of Biomass Conversion.* Open University Press, pp. 278.

Weberling, F. (1965) Typology of inflorescences. *Bot. J. Linn. Soc.* 59, 215–221.

Wickens, G.E. (1976) The flora of Jebel Marra (Sudan Republic) and its geographic affinities. *Roy. Bot. Gard. Kew. Bull. Addit. Ser.* HMSO, pp. 368.

Wilford, G.E. and Brown, P.J. (1994) Maps of late Mesozoic–Cenozoic Gondwana break-up: some palaeogeographic implications. pp. 5–13 in Hill, R.S. (Ed.) *History of the Australian Vegetation: Cretaceous to Recent.* Cambridge University Press, pp. 433.

Wilhelm, W.W. and McMaster, G.S. (1995) Importance of the phyllocron in studying development and growth in grasses. *Crop Sci.* 35, 1–43.

Wilkinson, M.J. and Stace, C.A. (1991) A new taxonomic treatment of the *Festuca ovina* L. aggregate (Poaceae) in the British Isles. *Bot. J. Linn. Soc.* 106, 347–397.

Woods, R.W. (1970) The avian ecology of a tussock island in the Falkland Islands. *Ibis* 112, 15–24.

Wu, S.H. and Tsai, L.-K. (1963) Cytological observations on the F_1 hybrid (*Oryza sativa* L. × *Pennisetum* sp.). *Acta. Bot. Sin.* 11, 293–307.

Yamura, A. (1933) Karylogische und embryologische studien über einige Bambusa. *Arten Botanical Magazine Tokyo* 47, 551–555.

Zambryski, P. (1983) Ti plasmid vector for the introduction of DNA into plant cells without alteration of their normal regeneration capacity. *EMBO J.* 2, 2143–2150.

Zevallos, M.C., Galinat, W.C., Lathrap, D.W., Lang, E.R., Marcos, J.G. and Klumpp, K.M. (1977) The San Pablo corn kernel and its friends. *Science* 196, 385–389.

Zhong, H., Bolyard, M.G., Srinivan, C. and Stricklen, M.B. (1993) Transgenetic plants of turf grass (*Agrostis palustris* Huds.) from micro projectile bombardment of embryonic callus. *Pl. Cell Repts.* 13, 1–6.

Zhou, G.-Y., Zxen, Y. and Yang, W. (1981) The molecular basis of remote hybridisation. An evidence for the possible integration of sorghum DNA into the rice genome. *Sin. Sci.* 24, 701–709.

Zubakov, V.A. and Borzenkova, I.T. (1990) Global palaeoclimate of the late Cenozoic. *Developments in Palaeontology and Stratigraphy. No. 12.* Elsevier, Amsterdam.

A Critical Glossary
of the Grasses

In its original form this glossary was first published in Chapman and Peat (1992). In the present version it is now substantially augmented.

The treatment here is to interest as much as to define. Some terms, even those long in use, yield thought-provoking ambiguities when carefully reconsidered.

Are botanical terms descriptive or analytic? They can be either. A leaf with a surface bloom can be described as glaucous and that is uncontroversial. If a leaf belonging to a grass is referred to as a 'phyllode', what seemed to be leaf is now regarded as an expanded petiole. The approach is 'analytic'. Nor does the matter rest there, since this view advocated by Arber (1918) is unhesitatingly rejected by Tomlinson (1970), with whom, on this point, present-day agrostologists tend to agree.

Again, some terms are more or less exclusive to grasses or have been extended from them subsequently. These include 'culm' and 'tiller' and seem to bear the stamp of practicality. Next there are those seemingly botanical terms, used for grasses in a way that properly surprises other botanists. They, assuming a pedicel supports a floret, find in grasses that it supports a spikelet. Moreover, what is a grass inflorescence? Do we mean the panicle or its component spikelets or both and are what we call 'racemes' condensed panicles? Reassuringly, some botanical terms are used for grasses no differently than elsewhere.

The approach adopted here is discursive where necessary. Etymology, surprisingly, is of relatively little use. While it tells us what a word means, that is of little help when we want to know how it is used. Terms such as 'anemophily' (noun) converted to 'anemophilous' (adjective) are defined using the former except where no such precedent exists as, for example, 'adventitious', 'iterauctant' and 'plicate'.

Not all the terms defined here occur in the present text but are intended to help with grass literature generally. 'Alicole' is an example. See Moss and Heslop-Harrison (1960) referred to in chapter 12.

A continuing source of interest is the relation to bamboos to the grasses generally. In most cases the definitions here are general to the Poaceae and are printed in Serif type. For convenience those peculiar to bamboos are printed in Sans Serif type.

Abaxial/Adaxial: Strictly, this refers to whether the surface faces away from or toward the axis of the originating primordium, i.e. outer or inner face. Commonly used for the lower and upper leaf surfaces respectively, but may also be applied to (say) lemma, palea or glumes. A vertical structure therefore has abaxial and adaxial surfaces.

Adventitious: Applied to roots arising from stem nodes to distinguish them from those originating at or from the primary root emerging from the germinating seed. cf. *Seminal*.

Agamospermy: See *Apomixis*.

Aleurone: One or more layers of cells formed from the outermost cells of the endosperm which store substantial quantities of protein. cf. *Bran, Grist*.

Alicole: Applied to maize and teosinte by Collins (1919), it refers to the spikelet or spikelets whether staminate or pistillate that are borne at a single point on the rachis considered as the axil, a point of attachment of a reduced branch. cf. *Pedicel, Sessile spikelet*.

Andromonoecy: Hermaphrodite and male flowers on the same inflorescence, as in many panicoid grasses.

Anemophily: Wind-mediated pollination.

Annual: Completing the life cycle within a year. cf. *Ephemeral, Perennial*.

Anthesis: Pollen dehiscence due to opening of the anthers. In functionally cleistogamous plants it precedes any flower opening. cf. *Cleistogamy, Protandry, Protogyny*.

Anthoecium (Anthecium): Although McClure and Söderstrom (1972) used the term to refer collectively to the lemma and palea, the earlier usage of Stapf (1904, 1926) that it included lemma, palea *and* flower is preferred.

Antipodal: Applies to nuclei and/or cells near the chalaza within the embryo sac. In grasses their number and the amount of DNA per nucleus are variable. cf. *Egg apparatus*.

Apomixis: Commonly used now to describe non-sexual reproduction via the seed. cf. *Apospory, Diplospory, Pseudogamy*. (The literature is not consistent; formerly apomixis meant all forms of vegetative reproduction including tillers, rhizomes, etc. The term 'agamospermy' was used for vegetative reproduction via the seed as a form of apomixis.)

Apomorph: A term used in an evolutionary sense to describe a supposedly derived character. cf. *Plesiomorph*.

Apospory: Accounts in grasses for most species where apomixis occurs, and is the development of an unreduced embryo sac from a somatic cell of the ovule in the nucellus, but can occur in the integuments. cf. *Apomixis, Diplospory, Pseudogamy*.

Arm cell: Specialised leaf cells, characteristic of Bambuseae and typically 'm'-shaped. cf. Fusoid cell.

Auricle: Extensions to the leaf lamina or, in bamboos, the leaf sheath projecting toward and sometimes enclosing the stem. Well known in wheat. cf. *Oral setae*.

Awn: A stiff projection usually from the lemma or glume either from the tip or the abaxial surface. If bent like a knee the awn is 'geniculate'. Awns that hydrate are called 'hygroscopic' and those that do not 'passive'.

Baccate fruit: The pericarp (the caryopsis wall) is fleshy in some bamboos such as *Melocanna*.

Bagasse: The residue of sugar cane stem after rollers have expressed the juice.

Boot: The flag leaf as it encloses the inflorescences before ear emergence. cf. *Flag leaf*.

Bract/Bracteole: A (sometimes leaf-like) structure subtending an inflorescence while a bracteole or bractlet subtends a flower. Since the spikelet in its various forms and the inflorescence branching system seem to have been repeatedly 'condensed', the bract or bracteole status of glumes, lemma and palea is dubious. A gemiparous bract is one having a bud in its axil.

Braird: A stand of newly emerged seedlings.

Bran: Husk of grain separated from the flour after grinding. In wheat, bran includes pericarp, integuments and nucellus derivatives. cf. *Aleurone, Grist*.

Branch: An axillary bud developing either from the primary stem or one of its subsidiaries obviously above the ground. cf. *Rhizome, Stolon, Tiller*.

Branch complement: An axillary bud can, especially in bamboos, divide to produce subsidiary apices, a group of which develops to produce the characteristic cluster. cf. *Gremial*.

Bulb/Bulbil: 'Bulbous'-shaped structures occur in Poaceae, but true bulbs, formed from swollen leaf bases, can be confused with corms. *Melica bulbosa* is corm forming, for example. Ref. Burns (1946). cf. *Corm*.

Bulliform cell: Water-retentive bubble-shaped cells occurring in groups between leaf veins, poor in contents. Alterations in their shape caused by loss of water allows leaves to inroll into a more tubular shape, thus diminishing transpiration.

Bundle sheath: See *Mestome-sheath*.

C₃ grasses: Those species (commonly temperate) whose first detectable photosynthetic intermediate sugar precursor is the 3-carbon compound 3-phosphoglycerate.

C₄ grasses: Those species (commonly tropical) whose first detectable photosynthetic intermediate sugar precursor is a 4-carbon compound, either malate or aspartate.

Caespitose: Tufted.

Callus: Can refer to either the hard tip of a spikelet or sometimes of the floret which assists its penetration of the soil, or, of course, more widely to often undifferentiated tissue raised in *in vitro* culture.

Carpel: Ovule-bearing structure that either singly or with others provides ovary, style and stigma tissue (collectively the pistil). In grasses the remnants of three, or more commonly two, carpels are believed to form the ovary and 'share' a common ovule.

Caryopsis: The grass fruit, normally dry at maturity, consisting of a single seed within, and including the ovary. cf. *Baccate fruit, Pericarp*.

Centrifugal/Centripetal: Applied to chloroplasts in bundle sheath cells, which are concentrated away from or toward the xylem cells respectively.

Chaff: A colloquial term referring to dried floral or spikelet remnants associated with a mature caryopsis.

Chalaza: Area of the ovule at the opposite end to the micropyle.

Chaparral: Refers to the sclerophylous dense scrub that forms on stony soils with little organic matter in the mediterranean region of California, and which resembles closely the 'maquis' and 'garrigue' sclerophylous woodland vegetation that occurs in the mediterranean regions of southern France and Spain as well as North Africa and Greece. cf. *Desert, Dune, Pampas, Prairie, Steppe, Veld*.

Chasmogamy: Refers to a flower which opens so as to permit both cross- and self-pollination. cf. *Anthesis, Cleistogamy*.

Chorology: The study of distribution and composition of elements in a flora (or fauna).

Cleistogamy: Refers to a flower which remains closed to permit only self-pollination (and by implication, self-fertilisation) although sometimes the flower opens later, as in wheat. cf. *Anthesis, Chasmogamy*.

Cleistogene: Modified spikelet containing self-compatible flowers within basal leaf sheaths.

Co-florescence: See *Florescence*.

Coleoptile: A sheath of tissue enclosing the shoot and from which leaves and the stem subsequently emerge. cf. *Coleorhiza, Plumule*.

Coleorhiza: A sheath of tissue enclosing the radicle prior to emergence. In monocots, the radicle is somewhat short-lived, the bulk of roots (adventitious) developing from the stem nodes.

Collar: The line of yellow tissue between sheath and blade on the abaxial surface of the leaf.

Condensation: A term used to describe loss of parts through evolution, particularly of the inflorescence which can alter its appearance to give a simpler, derived structure though one sometimes retaining hints of previous changes.

Conduplicate: Two parts folded together lengthwise as in emerging leaves of *Poa*. cf. *Involute*.

Connective: The band of tissue linking each anther locule, to which the distal tip of the filament is attached.

Corm: Enlarged, swollen stem tissue as in *Arrhenatherum elatius*. *Poa* curiously has both bulb-forming and corm-forming species, respectively *P. bulbosa* and *P. nodosa*. See Burns (1946). cf. *Bulb/Bulbil*.

Corn: A general name for barley, oats, rye and wheat in Britain. In America, corn refers to maize and the four crops mentioned earlier are referred to as 'small grains'.

Culm: Aerial stem of grasses. Normally vertical but can be prostrate or spreading. cf. *Tiller, Uniculm*.

Culm sheath: A leaf sheath wrapped around the culm with a diminutive leaf lamina.

$\delta^{13}C$: C_3 plants discriminate against $^{13}CO_2$ while C_4 plants do not. The organic matter of C_4 plants therefore is similar to that of the atmosphere in proportion of $^{12}C/^{13}C$. The ratio is determined by burning dried plant material and measuring the proportions of $^{13}CO_2$ and $^{12}CO_2$ in comparison with an internationally agreed standard. $\delta^{13}C$ is the proportionate change from this control expressed in parts per 1000. C_3 plants have lower values (-22 to $-35‰$) than C_4 (-9 to $-18‰$) and $\delta^{13}C$ is considered the most reliable way to distinguish between C_3 and C_4 types.

Decarboxylating enzymes: NAD-ME, nicotinamide adenine dinucleotidemalate enzyme; NADP-ME, nicotinamide adenine dinucleotide phosphate-malate enzyme; PEP/CK (or PCK) phosphenol pyruvate carboxykinase. These are the alternative enzymes which release CO_2 from C_4 acids in the bundle sheath cells of species with C_4 photosynthesis.

Desert: Technically, an environment so inhospitable that plant life is reduced to vanishing point. Antarctica provides an extreme case, more so than the Sahara which has a significant grass flora. cf. *Chaparral Dune, Pampas, Prairie, Steppe, Veld*.

Diaphragm: The sheet of tissue (woody in bamboos) separating the lumen of each internode in hollow-stemmed grasses.

Diaspore: Unit of dispersal. In cultivated wheat and maize the naked grain is separated from all the surrounding scales. In barley, oats and many other species, the palea and lemma are normally abscised with the grain (apart from naked grain types). In *Tristachya* and wild *Hordeum* spp., groups of three spikelets are shed. Numerous other variants occur. cf. *Caryopsis, Grain, Seed.*

Dichogamy: Male and female organs mature at different times permitting cross-pollination. cf. *Protandry, Protogyny.*

Dioecy: Male and female flowers borne on separate plants. cf. *Monoecy.*

Diplospory: An embryo sac developed from an unreduced megaspore mother cell. In grasses this is relatively rare but is reported in *Poa, Calamagrostis, Eragrostis* and *Tripsacum.* cf. *Apomixis, Apospory, Pseudogamy.*

Disjunct: One genus (or species) whose representatives are at separate locations. It does not follow automatically but it can sometimes be inferred that their distribution was once continuous.

Distichy: Parts arranged in two rows one each on opposite sides as in grass leaves and the florets within a spikelet. (Members of Cyperaceae have parts in threes thus providing a simple contrast with Poaceae.) cf. *Polystichy, Trimery.*

Dune: Elevated sand bank maintained by a combination of wind and psammophyte activity. cf. *Chaparral, Desert, Pampas, Prairie, Psammaphyte, Steppe, Veld.*

Ear: A general term for a cereal inflorescence which is compact or a spike. The female in ear in maize has protruding stigmas described as 'silks'. Seeds removed from the maize ear expose the axis called a 'cob'.

Egg apparatus: The association of egg cell with two synergids (in Poaceae) at the micropyle end of the embryo sac.

Elaiosome: Oil-bearing appendages sought by insects and facilitating seed dispersal. Occur in several grasses including *Yakirra.*

Embellum: A group of specialised transfer cells found associated with the micropyle in some grasses. See Busri *et al.* (1993).

Embryo sac: Within the ovule, the haploid megaspore (resulting from meiosis) undergoes three further sets of divisions to produce eight nuclei. These are rearranged as the egg apparatus, polar nuclei and antipodals. The embryo sac is the megagametophyte of flowering plants.

Endemic: Confined to a particular area. *Yakirra*, for example, is endemic to Australia. Endemism occurs in grasses but is not as important as in the palms, for example.

Endosperm: Tissue originating at double fertilisation from union of two polar nuclei with one male gamete. Flowering plants are unique in that the male parental genome contributes to the nutrient tissue for its offspring.

Ephemeral: Very short-lived: a plant completing its life cycle in appreciably less than 12 months. cf. *Annual, Perennial.*

Epiblast: A projection from the embryonic shoot opposite the scutellum. If monocots are assumed to be derived from dicots, the epiblast can be interpreted as a vestigial second cotyledon. Alternatively, the epiblast can be interpreted as a projection of the (stem-encircling) attachment of the single cotyledon, the scutellum. The epiblast is small in *Avena* and *Triticum* and larger in *Stipa* and *Oryza*, and in some grasses such as *Pennisetum* is absent.

Extravaginal: A branch which penetrates the base of its subtending leaf. cf. *Intravaginal.*

Fascicle: Clusters of spikelets found at intervals along the inflorescence branches of some bamboos, such as *Bambusa* for example.

Flag leaf: The last (and subtending) leaf before inflorescence emergence. As the largest and best illuminated leaf, its photosynthetic capacity can appreciably influence grain yield. If the sheath is enlarged and partially encloses the inflorescence, it is referred to as a spatheole.

Florescence: The series or group of flowers at the end of a shoot.

Co-florescence: Any lateral florescence below the main one.

Paracladium: A shoot arising from the main axis immediately below the basal internode of the main florescence.

Synflorescence: The whole floral aggregation in a plant, i.e. the system of the main florescence with its co-florescences.

The concept of 'florescence' and its subdivisions is primarily for use with bamboos. Even here its adoption has been restricted. See Calderon and Söderstrom (1973) and Weberling (1965).

Fusoid cell: Large central cells in the leaves of Bambuseae, on either side of the vascular bundle and surrounded by 'arm cells' these latter typically 'm'-shaped.

Gemmiparous bract: See Bract/Bracteole.

Glume: Bracts at the base of a spikelet (usually two) which do not themselves contain florets and are thus 'sterile'. In bamboos several 'glumes' might be present, but these are of uncertain affinity. Transitional glumes occur in some bamboos grading into sterile lemmas.

Gluten: A protein in wheat flour that adds elasticity to dough. Genetically it was contributed to hexaploid (bread) wheat from *Triticum tauschii* (*Aegilops squarrosa*, *Patropyrum tauschii*). It functions by allowing the bubbles of CO_2 produced by yeast to be retained, thus allowing the bread to rise (or be leavened).

Grain: At its simplest, the grain is synonymous with the caryopsis as in free-threshing crops such as wheat. In barley, the caryopsis need not be 'naked'

and can adhere to the lemma and palea being shed as a unit. cf. *Caryopsis, Diaspore, Seed*.

Gregarious flowering: Applied commonly to bamboos when many individuals of a species flower collectively after a long interval. 'Mast' flowering is the term preferred by Janzen (1976) but refers to the same phenomenon.

Gremial: Growing in a pollard-like cluster – applied to the branch complement.

Grist: a mixture of grain used for milling or malting. cf. *Bran*.

Halophyte: A plant able to tolerate saline conditions. cf. *Psammophyte*.

Hay: Animal feed from grass or other plants harvested, ideally, at maximum nutritional value and stored dry. cf. *Silage*.

Heterofertilisation: An occurrence where the egg and central cells of one embryo sac are fertilised by male gametes originating in different pollen tubes.

Hilum: The scar on the caryopsis surface revealed by abscission of the funicle (seed stalk).

Holocarpy (Monocarpy): Applied primarily to bamboos that die after flowering and fruiting but is also applicable to annual grasses or to the individual flowering culms of perennial species.

Homology: Used in two senses. It can mean corresponding chromosomes arising from each parent that will pair at meiosis. Applied to morphology, if the lodicule of a grass is considered to correspond to a petal of (say) *Tradescantia*, this could imply a common origin.

Hypopeltate appendix: A projection of the scutellum approximately parallel to the radicle in a grass embryo.

Ideotype: A man-made design for a plant, intended to aid a breeder in producing an improved variety. Although often presented as a diagram it does of course imply certain physiological properties such as high harvest index, more nutritional grains or improved pest and disease resistance. Classic refs: Donald (1968); Bunting (1971).

Induration: Owing possibly to the influence of genes from *Tripsacum*, the hard cob of maize is said to be indurated.

Inflorescence: A collection of spikelets arranged on a common branching system. cf. *Florescence, Panicle, Raceme, Spike*.

Ingera (Enjora): A fermented paste of flour subsequently cooked and often eaten with peppered meat. An Ethiopian food prepared from t'ef (*Eragrostis tef*).

Integument: Layers of cells surrounding the ovule. Grass ovules are invested by two layers of tissue and are therefore bitegmic. In both panicoids and pooids the inner integument covers the ovule sufficiently completely to form the micropyle, while the outer integument is respectively less and more complete in covering the ovule. cf. *Bran*.

Internode: The interval on a stem, rhizome or between two nodes; typically conspicuous in grasses. cf. *Phytomer*.

Intravaginal: A branch which remains enclosed by its subtending leaf for a substantial period of development. cf. *Extravaginal*.

Involucre: In general botanical usage a whorl of small leaves or bracts standing close underneath a flower or flowers. In *Cenchrus* the involucre is formed from sterile branches.

Involute: Inrolled to include the adaxial surface. cf. *Revolute*.

Iterauctant: (Lat. *iteratus* – repeated). A term applied to a congested bamboo inflorescence where pseudospikelets occur producing successive orders or branching. cf. *Pseudospikelet, Semelauctant*.

Keeled: Ridged like the bottom of a boat. In grasses the palea has two keels. The Oryzeae are interesting here since the palea has only one keel.

Kranz and non-Kranz: These refer to whether or not the vascular bundles of transverse leaf sections possess photosynthetic bundle sheaths, indicating C_4 or C_3 types respectively.

Lamina: The extended flattened portion of the leaf as opposed to the sheath enclosing the stem.

Lemma: The outer bract which, together with the palea, encloses the flower. Of interest in its detailed variation and therefore taxonomic usefulness. In vestigial florets it is often the last surviving remnant.

Leptomorph: See *Monopodial*.

Ligule: The adaxial extension of leaf sheath at its junction with the lamina. An 'outer' ligule described for *Arundinaria* is not an outer rim of the normal ligule but a separate organ.

Lodicule: The likely equivalent in grasses of the 'petal' seen elsewhere. Typically three in bamboos, two elsewhere. Lodicules inflate at anthesis to separate palea and lemma thus opening the floret.

Lumen: A hollow space such as exists in a stem internode or a single non-living cell.

Macrohairs: Large hairs found on grass leaf and stem surfaces, but cf. *Microhairs*.

Meristem: A cluster of dividing cells comprising the root and stem apices. A meristem occurs at the base of grass leaves rendering them less damaged by animal grazing or lawn mowing.

Mesocotyl: An interpolated node in the grass seedling separating coleoptile and cotyledon.

Mestome-sheath: A sheath of tissue which, if present, directly surrounds the vascular bundle of a grass leaf. Such a leaf is coded XyMS +. If absent the leaf is coded XyMS –. In some cases, e.g. *Neurachne*, it is the inner layer of a double bundle sheath which is photosynthetic. Since a non-photosynthetic layer is then *not* interposed between the photosynthesising layer and xylem, this grass type, though having double-sheathed bundles, is referred to as XyMS – (Hattersley and Watson, 1992).

Metaxenia: The phenomenon whereby tissues *outside* of the embryo sac are influenced by the pollen source. This is known for date palm (*Phoenix*) and should not be confused with xenia, a situation known in *Zea*. cf. *Xenia*.

Microhairs: Small hairs conveniently seen on leaf surfaces, which are of characteristic shapes for different groups of grasses. For detail see Watson and Dallwitz (1988), and Chapman (1992b). cf. *Microhairs*.

Micropyle: The aperture formed where the integuments do not completely envelop the nucellus. Before fertilisation it provides a port of entry for the pollen tube, and at seed germination ingress for moisture.

Monoecy: Separate male and female inflorescences on the same plant – maize is the most familiar example. cf. *Dioecy*. (See Jones, 1934, where a simple genetic manipulation converts moneocy to dioecy.)

Monopodial: Rhizomes which run indefinitely, producing culms from lateral buds. The synonym 'leptomorph' is sometimes used. cf. *Sympodial*.

MS: See *Mestome-sheath*.

NAD-ME: Refers to nicotinamide adenine dinucleotide co-factor to malic enzyme, a key enzyme in one version of the C_4 pathway.

NADP-ME: Refers to nicotinamide adenine dinucleotide co-factor to malic enzyme also a key enzyme in one version of the C_4 pathway.

Neck: Constricted part at the base of segmented vegetative axes.

Nerve: Often used as a synonym for vein (or vascular bundle). The animal associations are hardly appropriate when applied to plants. *Vein* is to be preferred.

Nobilisation: A process described for *Saccharum* where *S. officinarum* × *S. spontaneum* provides a hybrid with an unreduced chromosome number from the female parent. See Bremer (1961).

Node: The junction between two internodes, conspicuous in grasses, from which leaves, adventitious roots and branches can arise. The successive changes in development of nodes along a stem are of interest. cf. *Phytomer*.

Nucellus: Ovule tissue internal to the integuments and surrounding the embryo sac. Normally transitory in seed development, one reference shows it massively enlarged in sorghum followed *in vitro* pollination with maize (Dhaliwal and King, 1978).

Operculum: The 'lid' covering the single germ pore of a grass pollen grain.

Oral setae: Conspicuous bristles on the sheath near its junction with the leaf. cf. *Auricle*.

Orthostichy: Arrangements of vascular bundles that occupy opposite sides of a stem. Note that they are out of phase.

Palea: Commonly a two keeled structure found betwen lemma and lodicules. See, however, text discussion of 'Prophyllum' and in this glossary.

Pampas: Applies to large areas of open grassland in Argentina and Uruguay dominated by bunch grasses in the moister eastern region and by short grasses and xerophytic shrubs in the drier southern and western regions. cf. *Chaparral, Desert, Dune, Prairie, Steppe, Veld*.

Panicle: A branching inflorescence which in many genera is lax but in some instances is tightly compacted. In bamboos, the term panicle is applied to an inflorescence with apparently two or more orders of branching. *Eragrostis tef* shows this range within one species. cf. *Raceme, Spike*.

Paracladium: See Florescence.

PCA: Primary carbon assimilation tissue

used to describe C_4 leaf mesophyll.

PCK, PEP/CK: Phosphoenolpyruvate carboxykinase, an enzyme that decarboxylates oxaloacetate and a key enzyme in one version of the C_4 pathway.

PCR: Photosynthetic carbon reduction tissue the site of Calvin–Benson C_3 photosynthcsis which in C_4 plants is confined to the parenchyma bundle sheath.

Pedicel: A stalk supporting a spikelet. cf. *Alicole*. Curiously, in other plant families the pedicel supports the flower arising on a peduncle.

Perennial: Surviving over several (or even many) seasons. Perhaps in extreme cases up to 6000 years (in *Phragmites australis*). cf. *Annual, Ephemeral*.

Pericarp: Nominally the carpel tissue comprised of epi-, meso- and endocarp that surrounds (in grasses) the single ovule. cf. *Baccate fruit, Testa*.

Petiole: The leaf stalk. See the opening discussion of this Glossary.

Phenetic: Describing the appearance rather than the genetic constitution of an organism.

Photorespiration: Detectable in C_3 grasses where ribulose bisphosphate decarboxylase combines with oxygen rather than CO_2, eventually forming glycolate which is removed by a respiratory process.

Photosynthesis: The fixation of carbon as sugar from CO_2 and water with oxygen resulting.

Phyllocron: The interval between similar growth stages of successive leaves on the same culm (Wilhelm and McMaster, 1995). For extensive treatment of the phyllocron see *Crop Science*, 1994, 35, 1–49. cf. *Plastochron*.

Phytomer: A unit of development, defined as a leaf, the subadjacent internode, the node and its lateral bud, with adventitious roots if present. Ref. Clark and Fisher (1987). cf. *Node*.

Plastochron: The interval in time between the initiation of one primordium and the next. cf. *Phyllocron*.

Plesiomorph: A term used to describe a supposed primitive character. cf. *Apomorph*.

Plicate: Folded as a fan or approaching this condition.

Plumule: The embryonic shoot enclosed within, and later as it emerges from, the seed. cf. *Coleoptile*.

Polycross: Random mating between members of a group of selected genotypes by allowing open pollination among them.

Polygamy: Refers to one plant having many sexual partners but the preferred term is panmixis.

Polyphyletic: An individual taxon to which different (by implication very different) ancestors have contributed. cf. *Taxon*.

Polystichy: Among grasses, a term reserved for multiple (vertical) grain rows in maize. cf. *Distichy, Trimery*.

Prairie: Most often applied to grassland dominated by tall sod-forming grasses with 100% cover that is found covering extensive areas of North America. cf. *Chaparral, Desert, Dune, Pampas, Steppe, Veld*.

Prolifery: The preferred alternative to vivipary; it is applied to an inflorescence which has become vegetative.

Prophyllum/Prophyll: Perhaps the most contentious term in agrostology. Jackson (1928) regards it as equivalent to a palea. McClure (1966) on p. 93 calls it the first foliar appendage subtending a pseudospikelet but on p. 312 (glossary) defines it as a sheathing organ found in vegetative and inflorescence branches. Clayton and Renvoize (1986) equate the palea with a diminished prophyll of an axillary branch. It is also used sometimes in reference to unexpanded tiller buds. Turpin (1819) apparently coined the term setting up what Tomlinson (1970) called a 'wild goose chase'.

Blaser (1944) after critical scrutiny regards the prophyllum as neither more or less than a leaf, albeit sometimes modified. For discussion see the present text.

Protandry: Pollen shedding (anthesis) ahead of stigma receptivity. Conspicuous in many grasses.

Protogyny: Stigmas of a flower receptive ahead of pollen shedding (anthesis). Conspicuous in pearl millet. cf. *Dichogamy*.

Psammophyte: A plant, not necessarily a grass, found on and adapted to sandy habitats. cf. *Dune, Halophyte*.

Pseudogamy: A variant of double fertilisation where one male gamete fuses with the central cell and one male gamete approaches but does not fertilise the egg. A contributory factor in some apomicts. cf. *Apomixis*.

Pseudospikelet: A structure found in some bamboos. If the inflorescence is indeterminate each rachis branch ends in a spikelet but at its base is a short rachis clothed with lemma-like bracts and containing a meristem that will develop and end in a spikelet. The whole process can be repeated as for example in Bambusa multiplex. Ref. McClure (1966). Clayton (1990) drew attention to the multibranched stem of bamboo in relation to the bamboo branching panicle. The pseudospikelet remains a useful descriptive term. cf. Iterauctant, Semelauctant.

Pulvinus: A swelling in the axil of an inflorescence branch. More common in tropical grasses.

Raceme: Strictly, a raceme is an unbranched inflorescence with flowers borne on pedicels. Agrostologists have adopted the following procedure. They have utilised (for this purpose) the spikelet in place of the flower even though it is more correct to regard a spikelet as an inflorescence. Although therefore *Hordeum* for example is described as having a raceme-type inflorescence this should be understood in a loose descriptive sense rather than a strictly analytical one. In similar vein *Eragrostis* could be said to have digitately arranged racemes and *Spartina* as racemosely arranged racemes. In bamboos the term raceme is applied to an inflorescence with apparently one order of branching and clearly pedicellate spikelets. cf. *Panicle, Pedicel*.

Rachilla/Rhachilla: Subdivisions or branchlets of the rachis that support individual flowers, within the spikelet. cf. *Pedicel, Rachis*.

Rachis/Rhachis: The structure forming the axis or axes of the inflorescence. In cultivated cereals the structure is 'non-shattering'. cf. *Rachilla*.

Ramassage: The harvesting of wild grasses.

Reed: A plant growing in or near marshy conditions (a 'helophyte') having woody cane-like persistent stems.

Revolute: Inrolled to include the abaxial surface. cf. *Involute*.

Rhizanthogenes: Highly modified spikelets on underground stems (rhizomes). For example, *Eremitis*. cf. *Cleistogene*.

Rhizome: Stems developing below ground and bearing scale leaves from whose axillary buds ascending stems can arise. cf. *Branch, Stolon, Tiller*.

Root: In grasses, as in most monocots, the primary root system (sometimes called 'seminal', i.e. originating from the seed), is relatively short-lived, being replaced by adventitious roots arising from shoot nodes.

Rostrum: A cylindrical thickening of the lemma below the base of the awn found in some American *Stipa* for example. cf. *Stipe*.

Rubisco: Ribulose-1,5 bisphosphate carboxylase/oxygenase is the photosynthetic enzyme found in autotrophic plants. Ribulose 1,5 bisphosphate with CO_2 in the presence of Rubisco yields 3-

phosphoglyceric acid. If O_2 is substituted for CO_2, Rubisco catalyses the production of phosphoglyceric and phosphoglycollic acids.

Savanna: Grassland developing in semi-arid regions of the tropics and subtropics.

Scale leaf: A dry rudimentary or diminished version of larger greener leaves seen elsewhere on the same plant.

Scandent: Climbing, without the aid of tendrils or, according to Jackson (1928), climbing in whatever manner.

S-cleft: Space separating scutellum from coleorhiza.

Sclerenchyma: Lignified cells without protoplasts at maturity.

Sclerenchymatous girder: In cross-section an I-shaped structure comprised of a vascular bundle with sclerenchymatous cells toward the adaxial and abaxial surfaces (leaf-strengthening tissue).

Scutellum: The single absorptive cotyledon, abutting the endosperm. This latter is evanescent in *Melocalamus* and *Dinochloa* and the scutellum is consequently enlarged.

Seed: In every day usage in grasses is synonymous with the caryopsis although this is inaccurate. cf. *Diaspore, Grain*.

Semelauctant: A term applied to simple panicles where pseudospikelets do not congest the inflorescence. cf. *Pseudospikelet*.

Seminal: Arising from the seed. Commonly used for the seedling root system. cf. *Adventitious*.

Sessile: Literally, 'seated', that is arranged upon a supporting structure without a detectable stalk.

Sessile spikelet: One of the pair normally found in andropogonoid grasses, or a spikelet directly attached to the rachis, as in *Lolium* or *Triticum*. cf. *Alicole, Pedicel*.

Sheath: Normally applied to that part of the leaf enclosing the stem though seen impressively in the husks surrounding a maize ear.

Silage: Additional feed from grasses or other plants harvested, ideally, at maximum nutritional value and stored wet after anaerobic fermentation. cf. *Hay*.

Silica body: Silicon dioxide is laid down in appreciable amounts in many grass leaves, impeding section cutting for example, or offering some resistance to grazing. The crystals form characteristic shapes.

Silica cells: Those cells in which silica is deposited and which subsequently cease to be metabolically active.

Sinus: The depression between two lobes or teeth. Applied for example to the lemma tip in relation to the position of the awn.

Spatheole: A leaf-like structure enclosing, or at least subtending, a grass inflorescence. In cereals the term flag leaf is the commoner alternative. See Clayton and Renvoize (1986) p. 354.

Spike: An imprecise term applied to inflorescences with sessile spikelets, or sometimes to compact inflorescences. In no way comparable in importance to the term 'spikelet'. cf. *Panicle, Raceme*.

Spikelet: The key reproductive structure in grasses where (usually) two glumes subtend a group of flowers. Variations in the spikelet form a principal part of grass taxonomy. cf. *Pseudospikelet*.

Spinifex: A confusing situation exists. The descriptive term 'spinifex' covers a group of Australian chloridoid grass genera (*Troidia, Plectrachne, Symplectrodia* and *Monodia*). *Spinifex* is a panicoid genus of four species found on tropical sand dunes.

Steppe: Most often applied to grassland dominated by short bunch grasses often interspersed with shrubs, and a ground cover that may be less than 100%, that occurs in the semi-arid regions of North and South America, Africa, the Near East and Eurasia. cf. *Chaparral, Desert, Dune, Pampas, Prairie, Veld*.

Sterilisation (within the spikelet): An 'evolutionary' concept whereby once-functional flowers are thought to have become degenerate. 'Downward sterilisation' is an important characteristic of pooid grasses, while 'upward sterilisation' is found in panicoids.

Stipe: A stalk, but applied for example in *Microlaena stipoides* to that between the diminutive glumes and the florets. The genus *Stipa* apparently alludes to the stalked awns. cf. *Rostrum*.

Stolon: Horizontal stems produced above ground, rooting at some nodes. cf. *Branch, Rhizome, Tiller*.

Suite: Among alternative forms of C_4 photosynthesis it is possible to associate particular biochemical pathways with certain leaf structures. Each is a 'suite' of characters.

Sward: A growth habit maintained by rhizomes or stolons where the shoots, perhaps grazed, form a layer over the ground. cf. *Tussock*.

Sympodial: Applied to rhizomes where the apex turns upward to produce a culm. The superfluous synonym 'pachymorph' is sometimes used. cf. *Monopodial*.

Synergids: The two cells (in Poaceae) associated with the egg cell and collectively comprising the egg apparatus. One synergid is functional – the 'degenerating synergid' into which the pollen tube enters and discharges.

Synflorescence: See *Florescence*.

Syphonogamy: The delivery of gametes by a tube – in flowering plants the pollen tube which at its terminus enters the synergid to discharge.

SZ: The two-gene, multiple-allele incompatibility system, governing pollen–stigma interaction and apparently exclusive to the grasses.

Tabashir: Silica deposited within the culms of some bamboo species.

Taxon: A useful 'catch-all' term that, depending on context, can refer to family, genus or species, i.e. a taxonomic group.

Terete: Circular in transverse section.

Testa: The matured outer layer of the seed, however it is derived. Normally it consists of the two ovule integuments, but in most grasses its close association with the pericarp complicates delineation. *Sporobolus* is interesting because the pericarp extrudes the seed.

Tiller: A subsidiary culm arising at or near the base of the primary culm or one of its earlier subsidiaries. The grass habit is more 'bunched' (caespitose) if the rhizome or stolon connecting the tiller to its origin is shorter. cf. *Branch, Rhizome, Stolon*.

Trimery: Arranged in threes. In grasses the floral parts show modifications of this, being more strongly trimerous in bamboos, less so elsewhere. cf. *Distichy, Polystichy*.

Tussock: A growth habit where a collection of shoot apices forms a dome above the surrounding soil (or water as sometimes in the case of *Molinia*, for example). cf. *Sward*.

Ungulates: A general term referring to hoofed, chiefly herbivorous, mammals. Most ungulates belong to the orders Perissodactyla (odd- toed hoofed mammals such as the horse and rhinoceros) and Artiodactyla (even- toed mammals such as cattle, sheep and pigs). The order Ungulata, from which the term derives, is obsolete.

Uniculm: A growth habit where the aerial part of the plant is confined to a single (fruit-bearing) shoot.

Utricle: A modified leaf base found, for example, in *Coix*.

Valvule: A largely discarded term used to describe the palea. There is no case for its retention.

Vegetative reproduction: Nowadays used to include rhizomatous, stoloniferous and tiller multiplication. cf. *Apomixis*.

Vein: The visible strand running

(usually) the length of a stem or leaf, representing a vascular bundle and associated strengthening tissues. cf. *Nerve*.

Veld: Applies to grassland areas with little or no admixture of trees and shrubs that occur in the Karoo region of South Africa and the Kalahari region of Botswana. cf. *Chaparral, Desert, Dune, Pampas, Prairie, Steppe*.

Vernalisation: A necessary precursor involving cold treatment for the formation of reproductive primordia. Without it, a true winter cereal, sown in spring, would remain vegetative.

Versatile: Applied to anthers where the 'connective' between the anther lobes pivots on the filament thereby assisting pollen shedding.

Vivipary: See *Prolifery*.

Wolf plant: A characteristic growth habit described from *Agropyron desertorum* where the inflorescence long persists and causes avoidance of grazing. See Romo (1994).

Xenia: Known and well documented in *Zea*, it refers to the ability of the pollen source to affect endosperm colour or flavour. cf. *Metaxenia*.

XyMS: See *Mestome-sheath*.

Author Index

Taxonomic Index

Note: A problem for non-taxonomists is that of names for groups of genera, many of them now being superseded. For an indication of how these developed see Table 6.1, page 91. For a summary of contemporary usage the reader should consult either:

Clayton, W.D. and Renvoize, S.A. (1992) A system of classification for the grasses. pp. 338–353 in Chapman, G.P. (Ed.). *Grass Evolution and Domestication*. Cambridge University Press, pp. 390

or

Watson, L. (1990) World grass genera. pp. 258–265 in Chapman, G.P. (Ed.). *Reproductive Versatility in the Grasses*. Cambridge University Press, pp. 296.

Poaceae

Achlaena 129
Achnatherum splendens 168
Aegilops 91
    geniculata 88, 158
    squarrosa 242
Agenium 129
× *Agropogon* 76
Agropyron 40, 55, 112
    cristatum 87
    desertorum 249
    elongatum 169
    funeceiforme 48
    repens 28, 48, 121, 122, 142, 177
Agrosteae 99
Agrostideae 91
Agrostis 21, 32
    alba 191

    canina 11
    capillaris 122
    elatius 121
    magellarica 121, 122
    palustris 191
    stolonifera 11
    tenuis 11, 75
Aira 91
    caryophyllea 177
Alloteropsis 119
    eckloniana 119
    semi-alata 119, 132–3
Alopecurus 21, 53, 66, 91
    australis 121
    geniculatus 122, 150
    myosuroides 149–152
    pratensis 149

setigerus 28
Centotheca lappacea 21, 23
Centotheceae 34, 91
Centothecoideae 21, 41, 91, 95
Centropodia 131
Cephalostachyum 97
Chasmanthum 32
Chimonobambusa 97
    quadrangularis 19
Chimonocalamus 97
Chlorideae 91
Chloridinae 163
Chloridoideae 25–27, 34, 41, 53, 76,
    91, 119, 129, 131, 135, 136,
    153, 163
Chloridoidees 91
Chloris 32, 129, 133, 168
    gayana 25, 77, 164, 172, 173
    roxburghiana 161
Chrysopogon 174
    aucheri 72
Chusquea 3, 15, 44, 50
Cinna 91
Coix 53, 56, 90, 91, 248
    lachryma-jobi 28
Coleanthus 66
Cornucopiae 53, 66, 91
Cortaderia 118–119
    jubata 119
    seloana 19, 119
Criciuma 125
Ctenium 32
Cymbopogon 7, 91
    afronardus 31
    citratus 31
    coloratus 31
    densiflorus 31
    distans 31
    georugii 31
    giganteus 31, 174
    jwarancusa 31
    lehasianus 31
    martinii 31, 101
    microstrachys 31
    nardus 31
    polyheuros 31
    proximus 31
    sennarensis 31

    shoenanthus 31
    winterianus 31
× *Cynochloris* 133
Cynodon 5, 125, 55, 65, 71, 91, 129,
    133
    dactylon 5, 8, 11, 28, 152–155,
        161, 163, 164, 166–167,
        169
    plectostachyus 160
Cynodonteae 163
Cynosurus 21, 91
    cristatus 48
    echinatus 177

Dactylis 71
    glomerata 28, 48, 88, 177
Dactyloctenium 129
    aegyptium 163, 174
    ctenoides 171
Danthonia 32
    spicata 32, 158
    tenuior 85, 86
Dendrocalamus 15, 41, 42, 57, 59, 89,
    90, 91, 97, 144
    giganteus 144
    membranaceus 97
    sinicus 97
    strictus 72, 144
Deschampsia 21
    antarctica 117, 120–122, 131
    caerulea 11
    caespitosa 122
    chapmani 121
    penicellata 121
Diandrolyra 125
Dichanthelium 153
Dichanthium 174
    sericeum 86
Dichelachne macrantha 86
Digitaria 55
    abyssinica 163
    californica 164
    ciliaris 163
    decumbens 8, 162
    exilis 28
    iburua 28, 184

Other Organisms

Subject Index

Note: Geographical areas are entered under 'regions' and cereals and other grasses under 'common names' except for maize which has its own entry.

267

seedy perennial 141
   modifications 142
selection
   conscious 181
   K strategy 148–149
   R strategy 148–149
   salt tolerance 172
   unconscious 181
self-incompatibility 5, 72–75, 143
'set-aside' 138, 150
sheath primordia 43
shoot 42
smuts 32, 33–34
sodic tolerance 169
soil structure 85, 169
species
   cryptic 146
   divergence 188
   identity 102
spike 54–55
spikelet 5, 56
   'all purpose' 56
   evolution 59–60
   fossil 104
   modification 68
   pedicellate 68–70
   sessile 68–70
spinifex 155
   (cf. *Spinifex*)
spiral phyllotaxy 19, 145
stamen 58–59
'St Anthony's Fire' 31
stone terracing 161
sub families 14–28, 131–133
substituted seeding 142
succession 159, 177

taxonomy 89–102
   computerised sorting 95
   criteria 94, 95
   'excluded middle' 95
   experimental 92
   keys 95

   modern systems 94–96
temperature boundaries 99
tenacity 166
teosinte 210–211
   perennial 213–215
terracing 161, 167
tertiary 116
tiller 51, 167
topsoil 12, 164
totipotency 79
tourism 10
'transposons' 187
tree cover 165
tribe 91
   delineation 99
'true grasses' 3

ultraviolet radiation 126, 175–176
uniculm 158
urbanisation 8–9, 164

vascular bundle 48–50, 130
Vavilov 195, 210
vernalisation 155, 158
viability
   intertribal crosses 176
   seed 174
vivipary 68, 158
volcanic eruptions 176–177, 178–179
vulnerable grasses 177

water table 165
weediness 7–8, 35
weeds 162–164
wilderness 10

yield 190, 196, 217

zygote 78–80